Cladistics

Cladistics

Second Edition

The Theory and Practice of Parsimony Analysis

Ian J. Kitching, Peter L. Forey, Christopher J. Humphries, and David M. Williams

Oxford   New York   Tokyo

OXFORD UNIVERSITY PRESS

1998

Oxford University Press, Great Clarendon Street, Oxford OX2 6DP

Oxford New York

Athens Auckland Bangkok Bogota Bombay
Buenos Aires Calcutta Cape Town Dar es Salaam
Delhi Florence Hong Kong Istanbul Karachi
Kuala Lumpur Madras Madrid Melbourne
Mexico City Nairobi Paris Singapore
Taipei Tokyo Toronto Warsaw

and associated companies in
Berlin Ibadan

Oxford is a trade mark of Oxford University Press

Published in the United States
by Oxford University Press Inc., New York

A catalogue record for this book is available from the British Library

Library of Congress Cataloging in Publication Data

Cladistics: the theory and practice of parsimony analysis / Ian J.
Kitching . . . [et al.]. – 2nd ed.
(Systematics Association publications; 11)
Includes bibliographical references and index.
1. Cladistic analysis.   I. Kitching, Ian J.   II. Series.
QH83.C486 1998   570'.1'2 – dc21   98-10352
ISBN 0 19 850139 0 (Hbk)
ISBN 0 19 850138 2 (Pbk)

Typeset by Technical Typesetting Ireland, Belfast
Printed in Great Britain by Biddles Ltd., Guildford & King's Lynn

This book is dedicated to the memory of Colin Patterson who shared with us his deep insight into the theory and practice of systematics through discussion, argument, writings and above all, friendship.

Preface

The predecessor of this book originated as course material for a workshop sponsored by the Systematics Association in 1991. At that time, the Keepers of The Natural History Museum Botany and Entomology Departments, Dr Stephen Blackmore and Prof. Laurence Mound respectively, realized that there would be a broader demand for this information and encouraged its publication. The Systematics Association agreed to issue the manual as part of their Systematics Association Publications series. The workshop proved to be highly successful and was run for the next five years, during which time it was attended by over 100 students, together with 40 staff members of The Natural History Museum. In a modified format, the course was also run twice at the University of Verona, Italy, and once at the University of Massachusetts, Amherst. It also formed the basis of a module of The Natural History Museum/Imperial College of Science, Technology and Medicine M.Sc. course, *Advanced Methods in Taxonomy and Biodiversity*. As a publication, *Cladistics: a practical course in systematics* proved even more successful. The original 1992 hardback run sold out rapidly and the book was subsequently reprinted three times in softback format (1993, 1996, 1997).

Over this period there have been many developments in cladistic theory and practice. As our knowledge has advanced, some areas have become less significant, such as the *a priori* determination of character polarity. Others, such as the phenetic methods used to analyse large molecular data sets have been superseded by efficient and fast algorithms for parsimony analysis. Yet other areas grew and gained in importance. Three-item statements analysis rated less than a page in the original manual but is now perhaps the most contentious area in systematics, generating new ideas and forcing the critical re-appraisal of much cherished ideas. The whole field of tree support and confidence statistics had barely begun in 1992 but has now produced a wide variety of measures. Finally, the original manual was criticized by some as disjointed, appearing to be no more than a series of lecture notes put together in a single cover. If we are honest, that is exactly what it was.

When we decided to prepare a second edition, we resolved to address these issues, yet still produce a book of similar size to the original. Thus, we were faced with some hard choices. First, we decided that this was to be a book about **cladistics**. We therefore excluded those areas that we did not consider to be part of cladistics, such as phenetic and maximum likelihood techniques. There are other books dedicated to both topics that the reader can consult if so inclined. We also removed the chapters on cladistic biogeography and converting cladograms into formal classifications, again because specialist texts on these topics are available. We also wanted to stress a unified approach to the assembly and analysis of data, irrespective of its source. Hence there are no chapters dealing expressly with fossils or with molecular sequence data.

Instead, we have organized the book into nine chapters, beginning with a discussion of basic principles and concepts. The next six chapters follow the sequence of events in a cladistic analysis. Chapter 2 concerns characters and coding, that is, how we proceed from observations of organisms to an alphanumeric data matrix. Chapter 3 deals with cladogram construction from the data matrix and cladogram rooting, together with the related topic of polarity determination. Character optimization and the effects of missing data are considered in Chapter 4. Chapter 5 addresses character fit and weighting, while Chapter 6 provides the first comprehensive overview of cladogram support and confidence statistics. Chapter 7 discusses methods of consensus analysis. The pros and cons of simultaneous versus partitioned analysis form the subject of Chapter 8, while Chapter 9 discusses three-item statements analysis, a recently developed method that unfortunately has been the subject of much partisan and opaque writing. Finally, in response to numerous requests from students, we have included a glossary.

This book is the collective responsibility of all four authors. However, preparation of the first drafts of the text was undertaken as follows. Peter Forey wrote the sections on basic concepts, missing values and simultaneous versus partitioned analysis. Chris Humphries wrote the sections on characters, coding and consensus trees. Ian Kitching wrote the sections on cladogram construction, character polarity and rooting, optimization, confidence and support statistics and the glossary. David Williams wrote the sections on basic measures of fit, weighting and three-item statements analysis. Peter Forey prepared the figures, with considerable and capable assistance from his daughter, Kim. Finally, Ian Kitching undertook the role of editor, with the unenviable job of trying to marry the various sections into a single coherent and seamless whole. Any trivial mistakes that remain can be laid at his door but more fundamental disagreements should be taken up with all four of us. The choice of weapons will be ours.

Many individuals contributed greatly to this book. In particular, we would like to thank Gary Nelson, Mark Siddall, Karen Sidwell, Darrell Siebert and Dick Vane-Wright, who critically read through parts of the manuscript, and particularly to Andrew Smith and an anonymous reviewer, who read it all. We also thank Andrew Smith for permission to use his illustrations of the PTP test, Bremer support and the bootstrap and to Springer International for permission to reproduce the illustration of the 5S rRNA molecule of *Pedinomonas minor*. We also thank, most wholeheartedly, the Systematics Association for their continued support of this project.

I.J.K.
P.L.F.
C.J.H.
D.M.W.

London
September 1997

Contents

Authors

Peter L. Forey
Department of Palaeontology, The Natural History Museum, London

Christopher J. Humphries
Department of Botany, The Natural History Museum, London

Ian J. Kitching
Department of Entomology, The Natural History Museum, London

David M. Williams
Department of Botany, The Natural History Museum, London

1.
Introduction to cladistic concepts

1.1 DEFINITION OF RELATIONSHIP

Cladistics is a method of classification that groups taxa hierarchically into discrete sets and subsets. Cladistics can be used to organize any comparative data (e.g. linguistics) but its greatest application has been in the field of biological systematics. Cladistic methods were made explicit by the German entomologist, Willi Hennig (1950), and became widely known to English speakers in 1965 and 1966 under the name 'phylogenetic systematics'. Hennig wanted a method for implementing Darwin's concepts of ancestors and descendants. Hennig explained his ideas within an evolutionary framework; he wrote about species, speciation and the transformation of morphology through the process of evolution. In this introductory chapter, we will begin with Hennig's explanations but then slowly move towards the modern cladistic view, which dispenses with the need to rely on any particular theory of evolution for the analysis of systematic problems.

Hennig's most important contributions were to offer a precise definition of biological relationship and then to suggest how that relationship might be discovered. Formerly, the meaning of biological relationship was defined only vaguely. For example, a crab and lobster might be considered related because they show a high degree of overall similarity. Alternatively, these two animals might have been thought close relatives because it was easy to imagine how structures in one had transformed into structures in the other, perhaps through the link of a common ancestor. These definitions of relationship were absolute (X is related to Y).

Hennig's concept of relationship is relative and is illustrated in Fig. 1.1. Considering three taxa, the salmon and the lizard are more closely related to each other than either is to the shark. This is so because the salmon and the lizard share a common ancestor, 'x' (which lived at time t_2), that is not shared with the shark or any other taxon. Similarly, the shark is more closely related to the group salmon + lizard because the shark, salmon and lizard together share a unique common ancestor, 'y', which lived at an earlier time (t_1). The salmon and lizard are called sister-groups, while the shark is the sister-group of the combined group salmon + lizard. Similarly, the lamprey is the sister-group of shark + salmon + lizard. The aim of cladistic analysis is to hypothesize the sister-group hierarchy and express the results in terms of branching diagrams. These diagrams are called cladograms, a reference to the fact that

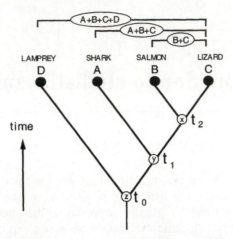

Fig. 1.1 Hennig's concept of relationship. For example, the lizard and the salmon are considered to be more closely related to each other than either is to the shark because they share a common ancestor, 'x' (which lived at time t_2), that is not shared with the shark or any other taxon.

they purport to express genealogical units or clades. The aim of cladistics is to establish sister-group relationships, and the concept of two taxa being more closely related to each other than either is to a third (the three-taxon statement) is fundamental to cladistics.

Sister-groups are hypothesized through the analysis of characters. These characters may be morphological, physiological, behavioural, ecological or molecular. The only requirement for cladistic analysis is that we translate what we observe into discrete characters, that is, a particular character is either present or absent; it is one colour or another; it is small or large; and so on.

1.2 TYPES OF CHARACTERS

Hennig made a distinction between two types of characters, or character states, and this distinction depended on where they occurred in the inferred phylogenetic history of a group. The character or state that occurs in the ancestral morphotype he called 'plesiomorphic' (near to the ancestral morphology) and the derived character or state he called 'apomorphic' (away from the ancestral morphology). The relation between character and character state is considered further in Chapter 2, where it will be seen that there are many methods for coding observations. Here, it is only necessary to emphasize that the terms apomorphic and plesiomorphic are relative terms—relative to a particular systematic problem. In Fig. 1.2a, character

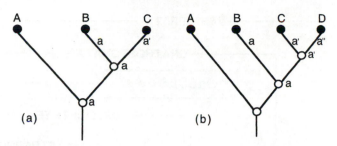

Fig. 1.2 Plesiomorphy and apomorphy are relative terms. (a) Character state 'a' is plesiomorphic and 'a″' is apomorphic. State 'a' is presumed to have been present in the ancestral morphotype that gave rise to taxa B and C. (b) Character state 'a″' is apomorphic with respect to 'a' but plesiomorphic with respect to 'a‴'.

state 'a' is plesiomorphic and 'a″' is apomorphic. State 'a' is presumed to have been present in the ancestral morphotype that gave rise to taxa B and C. In Fig. 1.2b, 'a″' is apomorphic with respect to 'a' but plesiomorphic with respect to 'a‴'.

Sister-groups are discovered by identifying apomorphic characters inferred to have originated in their most recent common ancestor and shared by its descendants. These shared apomorphies, or synapomorphies, can be thought of as evolutionary homologies, that is, as structures inherited from the most recent common ancestor. In Fig. 1.3, characters 3 and 4 are synapomorphies that suggest that the lizard and the salmon shared a unique common ancestor 'x'. The cladogram implies that characters 3 and 4 arose in ancestor 'x' and were then inherited by both the salmon and the lizard. In contrast, the shared possession of characters 1 and 2 by the salmon and lizard does not imply that they share a unique common ancestor because these attributes are also found in the shark. Such shared primitive characters (symplesiomorphies) are characters inherited from a more remote ancestor than the most recent common ancestor and are thus irrelevant to the problem of the relationship of the lizard and the salmon. However, with respect to the more inclusive three-taxon problem comprising the shark, salmon, and lizard, characters 1 and 2 are relevant. At this level, they are synapomorphies, suggesting that these three taxa form a group with a common ancestry at 'y'.

Alternatively, we may approach this problem by looking for the groups that are specified by different characters. In Fig. 1.3, given four taxa of unknown interrelationships, characters 1 and 2 suggest the group shark + salmon + lizard, while characters 3 and 4 suggest the group salmon + lizard. Furthermore, these four characters suggest two nested groups, one more inclusive than the other ((Shark (Salmon, Lizard)).

Here we recognize that synapomorphy and symplesiomorphy describe the status of characters relative to a particular problem. For example, characters 3 and 4 are synapomorphies when we are interested in the relationships of

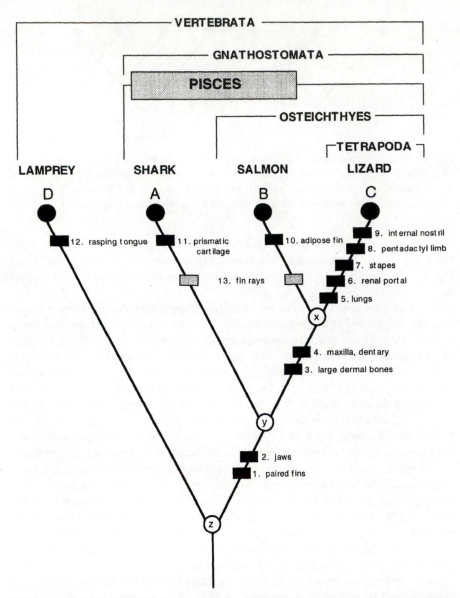

Fig. 1.3 Cladogram for the lamprey, shark, salmon and lizard. Monophyletic groups are established on the basis of synapomorphies (characters 1–4), while autapomorphies (characters 5–12) define terminal taxa. Character 13 conflicts with this hypothesis of relationships, suggesting instead a relationship between the shark and salmon. See Fig. 1.4 and text for further explanation.

the salmon, lizard, and shark, but symplesiomorphies if the problem involves the relationships of different species of lizards or different species of salmon.

Hennig recognized a third character type, which comprises those characters that are unique to one species or one group, such as characters 5–9 in the lizard, 10 in the salmon, 11 in the shark and 12 in the lamprey. These he called autapomorphies, which in Fig. 1.3 define the terminal taxa A–D. Autapomorphies can be thought of as the fingerprints of the terminal taxa.

The characters used to determine relationships are apomorphic characters or character-states and this implies acceptance of a theory of transformation (absence → presence, or condition a → a′). Hennig (1966) believed that there were several criteria by which we could recognize plesiomorphic and apomorphic states even before we started any cladistic analysis. The two most frequently employed today are the ontogenetic criterion and outgroup comparison. These are considered at length in Chapter 3, where we will also learn that the distinction between plesiomorphic and apomorphic resides in rooting the cladogram, that is, in the choice of the taxon that is to be the starting point for our theories of character evolution.

1.3 PARSIMONY

Cladistic analysis orders synapomorphies into a nested hierarchy by choosing the arrangement of taxa that accounts for the greatest number of characters in the simplest way. For instance, Fig. 1.3 accounts for most of the characters by assuming that each appeared only once in history, has been retained by all descendants and has never been lost. But usually the data will not all suggest the same groupings. For example, in Fig. 1.3, character 13 (fin rays) is uniquely shared by the shark and the salmon and suggests the grouping shark + salmon. In fact, this is a traditionally recognized formal Linnaean taxon, Pisces (fishes). Why therefore should we favour a different grouping? Consider two alternative solutions shown in Fig. 1.4. The cladogram shown in Fig. 1.4a recognizes the group Pisces (shark + salmon) based on one synapomorphy, character 13. In contrast, the cladogram in Fig. 1.4b recognizes the group Osteichthyes (salmon + lizard) based on two synapomorphies (characters 3 and 4). The second cladogram accounts for more characters as having arisen only once. If there are alternative solutions, then we would choose the simplest or most parsimonious pattern of character distribution. Parsimony is the universal criterion for choosing between alternative hypotheses of character distribution just as it is a universal criterion for choosing between any competing scientific hypotheses. It needs to be pointed out here that parsimony is simply the most robust criterion for choosing between solutions. It is not a statement about how evolution may or may not have taken place.

Parsimony is fundamental to cladistic analysis and may be explained in a slightly different way. Suppose we had six characters distributed among four taxa as shown in the taxon/character matrix in Fig. 1.5a. Taxon A has none

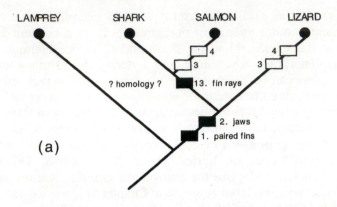

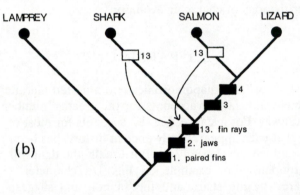

Fig. 1.4 Using parsimony to choose between two competing hypotheses of relationship. (a) The shark and the salmon form a monophyletic group based upon shared possession of fin rays (character 13). However, this topology requires us to hypothesize that characters 3 and 4 each arose independently in the salmon and the lizard. (b) Alternatively, the salmon and the lizard form a monophyletic group based upon characters 3 and 4, with character 13 now being considered homoplastic. This cladogram is preferred to that in (a) because it is more parsimonious. Character 13 may still be a synapomorphy but at a more inclusive level (albeit with some homoplasy) and is shown repositioned as such by the arrows.

of the characters but the other three taxa each have a different complement. Characters 2 and 4 are autapomorphies, since they are each present in only one of the taxa. They are uninformative for grouping taxa (they serve only to diagnose these terminal taxa). Characters 1, 3, 5 and 6 are potentially useful because they are present in more than one taxon. Given the three taxa that have potentially informative information, there are three ways in which we could arrange these taxa dichotomously (Fig. 1.5b–d).

If we now place each of the characters, according to the groups that they

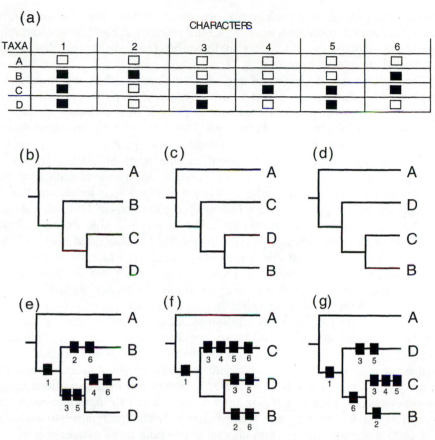

Fig. 1.5 Explanation of parsimony in terms of analysis of character distributions. (a) A data matrix of six characters (1–6) distributed among four taxa (A–D). Plesiomorphic states are indicated by open boxes, apomorphic states by solid boxes. (b–d) The three possible resolutions of taxa B–D relative to taxon A. (e) Placing the characters on the topology in (b) requires seven steps. Characters 1–5 appear only once while character 6 appears twice. This is the optimal, most parsimonious solution. (f) Placing the characters on the topology in (c) requires nine steps. Characters 1, 2 and 4 appear only once but characters 3, 5 and 6 all appear twice. This is a suboptimal solution. (g) Placing the characters on the topology in (d) requires eight steps. Characters 1, 2, 4 and 6 appear only once but characters 3 and 5 both appear twice. This is also a suboptimal solution.

specify, on each of these possible cladograms (Fig. 1.5e–g), then we obtain three different results. The cladogram in Fig. 1.5e shows that all but one of the characters appears only once. However, in this solution, we must assume that character 6 appears twice, once in taxon B and once in taxon C, which are not sister-groups.

We can do the same exercise for the cladograms in Fig. 1.5f and Fig. 1.5g.

However, in these two cladograms, we must assume that two or more characters appear more than once. Since the solution in Fig. 1.5e accounts for the distribution of the characters in the most economical way, this is the solution that we would prefer.

The distribution of characters can also be thought of as the number of steps on a cladogram. In Fig. 1.5, the number of steps is counted as the number of instances where a character is gained. In the cladogram in Fig. 1.5e, this is seven. The other cladograms (Figs 1.5f and 1.5g) are more costly, requiring 9 and 8 steps respectively. The idea of steps is a little more subtle because a single character may appear at one point on a cladogram and disappear again at another point. For example, another explanation of the distribution of character 6 on the cladogram in Fig. 1.5e is to assume that it was gained by the group B + C + D and then lost again in D. Each change, whether gain or loss, is considered to be a step. In this example, both accounts of character change demand two steps. Cladists often speak of the length of a cladogram and this is what they mean—the number of character changes, irrespective of whether those changes are gains or losses. The output from cladistic analyses using computer packages invariably gives cladogram lengths, as well as other statistics (see Chapter 5). Cladists also speak of the most parsimonious solution as being the optimal cladogram, and the other cladograms (that is, those requiring more than the minimum number of steps to explain the character distributions) as suboptimal.

It is possible that there are two or more equally parsimonious solutions for a set of characters. Then, we may prefer to accept one of the solutions based on other criteria, such as a closer agreement with the stratigraphic record or by differentially weighting one type of character change relative to another. For certain applications, we may choose to combine those elements common to the different solutions to make a consensus tree (see Chapters 7 and 8).

Any solution, or solutions, at which we arrive is a summation of the relationships among characters. Just as taxa may be related to one another, we can visualize characters as being related to one another. Fig. 1.6 illustrates the several ways in which characters can be related. Suppose that we have a character, 1, which is present in three taxa: A, B and C. This character can be said to specify a group A + B + C. Suppose that we now discover a second character, 2, which specifies a group D + E. Since no common taxa are involved the characters can be said to be consistent (strictly 'logically consistent') with one another. They can have no influence on one another because they specify completely different groups. When we add a third character, 3, that is present only in taxa B and C, this character specifies a subgroup of an original larger group A + B + C. This character 3 is also said to be consistent with character 1. Character 4 is found in taxa A, B and C and thus specifies exactly the same group as character 1. Characters 1 and 4 are said to be congruent with one another. Character 5 is found in taxon C and taxon D. Character 5 specifies a group C + D, which is totally unlike any group

(a)

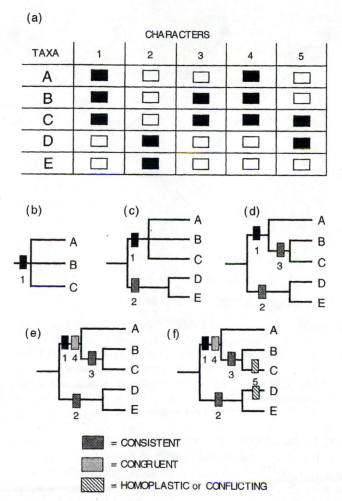

CHARACTERS

TAXA	1	2	3	4	5
A	■	☐	☐	■	☐
B	■	☐	■	■	☐
C	■	☐	■	■	■
D	☐	■	☐	☐	■
E	☐	■	☐	☐	☐

■ = CONSISTENT

▨ = CONGRUENT

▨ = HOMOPLASTIC or CONFLICTING

Fig. 1.6 Relationships among characters. (a) A data matrix of five characters (1–5) distributed among five taxa (A–E). Plesiomorphic states are indicated by open boxes, apomorphic states by solid boxes. (b) Character 1 is shared by three taxa, A, B and C, which form the initial group. (c) Character 2, present in taxa D and E, specifies a different group from character 1. (d) Character 3, shared by taxa B and C, specifies a subset of the initial group. (e) Character 4, present in taxa A, B and C, specifies the same group as character 1. (f) Character 5, present in taxa C and D, specifies a different group C + D that conflicts with the initial group. With regard to character 1, characters 2 and 3 are consistent, character 4 is congruent, while character 5 is in conflict.

specified by the other characters. In other words, the group that this character specifies conflicts with those specified by the other characters. Character 5 is said to be homoplastic.

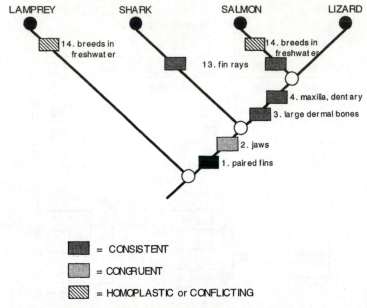

Fig. 1.7 Application of the character relationships shown in Fig. 1.6 to a cladogram
of the lamprey, shark, salmon and lizard. With reference to character 1, character 2 is
congruent because it specifies the same group. Characters 3 and 4 are consistent
because they specify a subgroup of that specified by character 1. Character 13 is
also consistent because it specifies a subgroup, even though that subgroup does not
appear in the most parsimonious solution. Character 14 conflicts with character 1
and is homoplastic.

Returning to the real example, we can recognize several types of character
interaction. In Fig. 1.7, and taking character 1 as the reference, character 2 is
congruent with it because it specifies the same group. Characters 3 and 4 are
consistent with characters 1 and 2 because they specify a subgroup of the
group specified by character 1. Character 13, which is shared between the
shark and the salmon, is also consistent with character 1 because it specifies a
subgroup, even though that subgroup does not appear in the most parsimonious solution. Character 14, which is shared between the lamprey and the
salmon, conflicts with character 1 and is thus homoplastic.

1.4 GROUPS

As a result of the relative definition of relationship, Hennig identified three
types of groups, which he recognized on the basis of ancestry and descent.
Using Fig. 1.8 as a reference the following groups may be recognized.

1. A monophyletic group contains the most recent common ancestor plus all
   and only all its descendants. In this figure, such groups would be ancestor
   'x' and salmon + lizard; or ancestor 'y' and shark + x + salmon + lizard; or

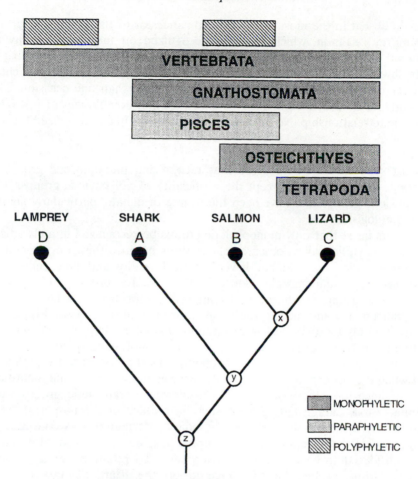

Fig. 1.8 The three types of groups recognized by Hennig on the basis of ancestry and descent.

ancestor 'z' and lamprey + y + shark + x + salmon + lizard. In this particular example, the monophyletic groups have formal Linnaean names: they are Osteichthyes, Gnathostomata and Vertebrata respectively.

2. A paraphyletic group is what remains after one or more parts of a monophyletic group have been removed. The group shark + salmon is a paraphyletic group that has been traditionally recognized as Pisces (fishes). However, one of the included members (the salmon) is inferred to be genealogically closer to the lizard, which is not part of the group Pisces. The shark and salmon share an ancestor (y) but not all the descendants of that ancestor are included in the group.

3. A polyphyletic group is defined on the basis of convergence, or on non-homologous, conflicting or homoplastic characters assumed to have

been absent in the most recent common ancestor of the group. The group lamprey + salmon, which might be recognized on the shared ability to breed in freshwater, would be considered a polyphyletic group. Breeding in freshwater in vertebrates might be considered to be an apomorphic character but this is inferred to have arisen on more than one occasion. The character by which we might recognize it is non-homologous: it is a false guide to relationship. No Linnaean taxon has ever been recognized for this group.

Most systematists would agree that recognizing monophyletic groups is desirable and would also accept the artificiality of polyphyletic groups. It is paraphyletic groups that have been the source of debate, particularly among palaeontologists.

Cladists insist that only monophyletic groups be recognized in a classification. Paraphyletic groups obscure relationships because they are not real in the same sense; they do not have historical reality and they cannot be recognized by synapomorphy alone. In Fig. 1.8, for instance, Pisces is a paraphyletic group recognized by having synapomorphies of a larger group (Gnathostomata) and lacking the synapomorphies of Tetrapoda. Fishes are distinctive only because they lack tetrapod characters. It turns out that the 'defining attributes' of paraphyletic groups are symplesiomorphies.

In the past, when evolutionary taxonomy was the classificatory paradigm and when the contribution of palaeontology was thought to be the identification of ancestors, there were two reasons why paraphyletic groups were popular. First, paraphyletic groups, such as reptiles or gymnosperms, were justified on the basis that extra evolutionary information was conveyed by distinguishing them from highly apomorphic relatives. That extra information was considered to be evolutionary divergence. To retain Pisces as a paraphyletic group in Fig. 1.3 and separate off the lizard (Tetrapoda) in a collateral group was done to emphasize the many autapomorphies (characters 5–9) of this latter group. In the terminology of evolutionary taxonomy, these tetrapod characters were evidence that tetrapods had shifted to a new adaptive zone (involving life on land, receiving stimuli through air rather than water, etc.). Similarly, retention of the paraphyletic group 'gymnosperms' was justified in order to recognize the evolutionary divergence of angiosperms. In a cladistic classification such divergence would be expressed by means of the number of autapomorphies identifiable in tetrapods or angiosperms. Conversely, to retain paraphyletic groups means that we lose cladogenetic information. Not all fishes are each others' closest relatives; some are genealogically more closely related to tetrapods than others.

Second, paraphyletic groups are popular and have been commonly recognized in palaeontology because traditionally they are the ancestral groupings (fishes ancestral to tetrapods, reptiles ancestral to birds and mammals, gymnosperms ancestral to angiosperms, algae ancestral to tracheophytes, and

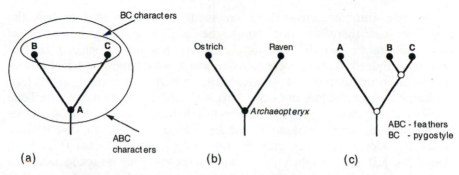

Fig. 1.9 Ancestors cannot be distinguished as individual taxa because they are wholly primitive with respect to their descendants and thus have no features by which they can be unequivocally recognized. (a) Three taxa, A, B and C, form a group because they share ABC characters. Taxa B and C are sister-groups because they share BC characters. If A is considered to be the ancestor of B and C, it can be placed at the origin of B and C only if it lacks any distinguishing characters of its own. Otherwise, it would be placed as the sister-group of B + C. In other words, ancestor A can only be recognized because it possesses ABC characters but lacks BC characters. (b) *Archaeopteryx*, equivalent to taxon A in (a), is the traditional ancestor of the birds. It has the synapomorphies (e.g. feathers) that are found in all birds, including the ostrich (taxon B) and the raven (taxon C), but lacks the synapomorphies of the ostrich + raven such as a pygostyle. In terms of character distribution, *Archaeopteryx* simply does not exist. (c) To circumvent this problem, cladists place ancestors as the sister-group to their putative descendants and accept that they must be nominal paraphyletic taxa.

so on). But all of these paraphyletic groupings are based on symplesiomorphies and can only be recognized by what they do not have: fishes do not have the autapomorphies of tetrapods, gymnosperms do not have the autapomorphies of angiosperms.

This should not really be a surprise because ancestors must, by definition, be wholly primitive with respect to their descendants. But what this also means is that they cannot be distinguished as individual taxa. Consider Fig. 1.9a, in which taxon A is ancestral to descendent taxa B and C. This diagram has been established on the distribution of characters, which are the only observable attributes of taxa. All taxa are grouped because they share ABC characters. Taxa B and C are sister-groups because they share BC characters. A can only be placed at the origin of B and C if it lacks any distinguishing characters of its own. Otherwise, it would be placed as the sister-group of B + C. In other words, A can only be recognized because it possesses ABC characters but lacks BC characters. Fig. 1.9b shows one traditionally recognized ancestor, *Archaeopteryx*. *Archaeopteryx* (equivalent to taxon A in Fig. 1.9a) has feathers, which is the most obvious synapomorphy of all birds, including the ostrich (taxon B) and the raven (taxon C). However, *Archaeopteryx* lacks the synapomorphies of the ostrich + raven such as a pygostyle. But, of course, there are many other animals that lack the pygostyle and therefore this cannot be a distinguishing character of *Archaeopteryx*. In fact, to date, *Archaeopteryx* has no recognized autapomorphies. Indeed, if there were, *Archaeopteryx* would have to be placed as the sister-group to the rest of the birds. In terms of unique characters, *Archaeopteryx* simply does not exist. This is absurd, for its remains have been excavated and studied. To circumvent this logical dilemma, cladists place likely ancestors on a cladogram as the sister-group to their putative descendants and accept that they must be nominal paraphyletic taxa (Fig. 1.9c). Ancestors, just like paraphyletic taxa in general, can only be recognized by a particular combination of characters that they have and characters that they do not have. The unique attribute of possible ancestors is the time at which they lived. After a cladistic analysis has been completed the cladogram may be reinterpreted as a tree (see below) and at this stage some palaeontologists may choose to recognize these paraphyletic taxa as ancestors, particularly when they do not overlap in time with their putative descendants (see Smith 1994a for a discussion). The logical impossibility of placing real taxa as ancestors in cladistic analysis has the further consequence that the ancestors 'x', 'y' and 'z', which have been placed on several of the figures, should be considered as *hypothetical* ancestors representing collections of characters.

Up to this point, groups have been described in Hennig's terms of common ancestry. But groups are not discovered in this way. In practice, they are discovered through analysis of character distributions. So we must return to characters to look again at the definition of groups. We can separate characters on their ability to describe groups. Those characters that allow us to specify monophyletic groups are synapomorphies. Monophyletic groups are discovered by finding synapomorphies. A very important conceptual leap came when homology was equated with synapomorphy (Patterson 1982). This has important consequences, for it means that homologies are hypotheses; hypotheses to be proposed, tested and, perhaps, falsified.

We can illustrate this by returning to Fig. 1.4. Let us assume that we have arrived at the hypothesis of relationships shown in Fig. 1.4a. This hypothesis recognizes that the salmon + shark is a monophyletic group, discovered by suggesting that the shared possession of character 13 (fin rays) is a synapomorphy or an homology. However, this hypothesis can be shown to be false because other synapomorphies suggest that the salmon and the lizard form a monophyletic group recognized by the shared possession of characters 3 and 4. The original hypothesis of homology has been tested and shown to be false. It may still be an homology but at a higher hierarchical level, as shown in Fig. 1.4b, where it specifies the larger group of shark + salmon + lizard. However, as an homology of the group shark + salmon, it is false.

So, an hypothesis of homology is tested by congruence with other characters. It should be obvious from this that homologies are continually being tested (the three tests of homology are discussed in Chapter 2). Discovery of

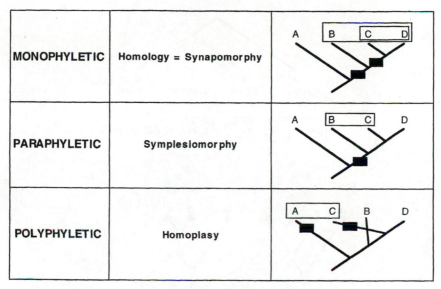

Fig. 1.10 The three types of groups recognized by Hennig defined in terms of character distributions. Monophyletic groups are discovered through homologies (synapomorphies); paraphyletic groups are those based upon symplesiomorphies; polyphyletic groups are founded upon homoplastic characters.

homology is at the heart of cladistic analysis. It is important to note, however, that there is no need to appeal to any specific theory of evolution in order to discover an homology. Evolutionary theory may help explain homology but it is unnecessary for discovering homology.

Monophyletic groups can be discovered through homologies and they are the only kind of group that can be justified by objective boundaries. Paraphyletic groups are those groups recognized by symplesiomorphies, that is, characters properly applicable at a more inclusive level in the hierarchy. Polyphyletic groups are recognized by homoplastic distributions of characters. These three relations are shown in Fig. 1.10.

1.5 CLADOGRAMS AND TREES

Throughout this chapter we have been slowly moving away from Hennig's evolutionary explanations for concepts of relationship, characters and groups. To conclude this chapter, we must make the important distinction between cladograms and trees.

The relationships between lamprey, shark, salmon and lizard, is drawn in Fig. 1.3 as a branching diagram—a cladogram. A cladogram has no implied time axis. It is simply a diagram that summarizes a pattern of character

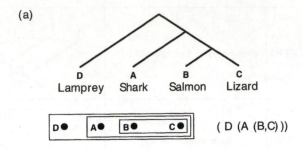

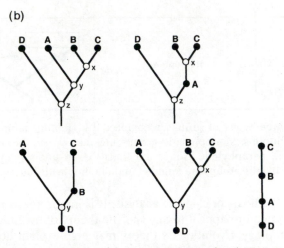

Fig. 1.11 Cladograms and trees. (a) the cladogram from Fig. 1.3 depicting the relationships of the lamprey, shark, salmon and lizard redrawn as a Venn diagram and in parenthetic notation. (b) Five of the 12 possible trees that can be derived from the cladogram in (a).

distribution. The nodes of the branching diagram denote a hierarchy of synapomorphies. There is no implication of ancestry and descent. It could just as easily be written in parenthetical notation or illustrated as a Venn diagram, as shown in Fig. 1.11a. The character information contained in the Venn diagram is compatible with a number of derivative evolutionary trees, which do include a time axis and embody the concepts of ancestry and descent with modification. Five such trees (out of a total of 12) are shown in Fig. 1.11b. Some of these trees assume that one or more of the taxa (A, B, C, D) are real ancestors. Other trees include hypothetical ancestors (x, y, z). Only one tree has the same topology as the cladogram and this is the one in which all the nodes represent hypothetical ancestors. The other trees contain one or more real ancestors. Choice among these trees depends on factors other than the distribution of characters over the sampled taxa, which is the only empirical content. Selection of one tree in preference to any other may depend on our willingness to regard one taxon as ancestral to others. Alternatively, we might say that some trees containing real ancestors are less likely to be true than others because of unfavourable stratigraphic sequences. The important point is that evolutionary trees are very precise statements of singular history but their precision is gained from criteria other than character distributions. These trees cannot be justified on characters alone.

The distinction between cladograms and trees is important because many people have taken the cladogram to be a statement about evolution. To do this we must be prepared to accept other beliefs, for example, that evolution is parsimonious or that evolution proceeds exclusively by branching. Many of the criticisms of cladistics are levelled at the claim that these are unrealistic assumptions of evolution. Indeed they are. But they are not assumptions of cladistics or cladograms. They are assumptions of trees. The cladogram, as a distribution of characters, is the starting point for further analysis. In practice, many systematists do turn their cladograms into trees in order to say something about evolution. Furthermore, they may incorporate evolutionary assumptions concerning the relative likelihood of character change, such as the impossibility of regaining a character once lost (Dollo's law), or the greater likelihood of nucleotide changes taking place within loop regions rather than stem regions of molecules. Some of these assumptions are considered further in Chapters 4 and 5.

1.6 TREE TERMINOLOGY

Having made the distinction between cladograms and trees, it needs to be stressed that, unfortunately, cladistic analysis often speaks only of trees, tree lengths and tree statistics. This is an historical legacy from Hennig's original formulation of phylogenetic systematics, as well as a mathematical convention for describing branching diagrams as trees.

There are several terms regarding trees that the reader will frequently meet. Trees have a root, which is the starting point or base of the tree. The branching points are called the internal nodes, while the segments between nodes are internodes or internal branches. Taxa placed at the tips are terminal taxa and the segments leading from internal nodes to a terminal taxon are called terminal branches.

Sometimes trees are drawn such that each of the branches is of equal length, irrespective of how many character changes may be assigned to it. This is called a non-metric tree. Another description is a metric tree in which the relative lengths of the branches are drawn to reflect the numbers of character changes, which may be different in different parts of the tree. The results of analyses of molecular data are often depicted as metric trees in order to emphasize the great variation in numbers of character changes that often occur on different branches. A third type of tree is the ultrametric tree in which each of the terminal taxa is fixed at the same distance from the root by assuming a constant molecular clock.

1.7 CHAPTER SUMMARY

1. The cladistic concept of relationship is relative. Taxon A is more closely related to taxon B than either is to a third taxon C or any other taxon.

2. Taxa A and B are sister-groups and this combined group will have its own sister-group. The aim of cladistic analysis is to discover sister-group relationships.

3. Sister-groups are discovered by plotting the distributions of synapomorphies.

4. Characters have relationships to one another on a given cladogram. They may be congruent, consistent or homoplastic.

5. Cladistic groups may be monophyletic, paraphyletic or polyphyletic. Only monophyletic groups are real.

6. Monophyletic groups are recognized by homology. Homology equals synapomorphy.

7. Cladograms are statements about character distribution. There may be several evolutionary trees compatible with one cladogram but most of these make additional assumptions beyond those of character distribution.

2.
Characters and character coding

2.1 INTRODUCTION

Cladistic analysis consists of three processes: discovery or selection of characters and taxa, coding of characters, and determination of cladograms that best explain the distribution of characters over the taxa. Although the three operations are inextricably interlinked, much of the literature deals with analysis of coded data matrices and affords less space to reviewing the principles of character discovery and coding from raw observations. In this chapter, we describe the interaction between discovery and coding, discuss different kinds of characters, and describe the methods and pitfalls associated with different coding procedures.

The operations of cladistic analysis are strongly influenced by the selection and resolution of taxa and characters. The kinds of data available (skeletons, soft parts, nucleotide sequences, etc.) inevitably affect choice of cladistic method. Conversely, the choice of analytical method may set bounds for the kinds of data that can be analysed and the format in which those data are recorded. Thus, the problem lies in defining which data will be recorded, how those data are coded and whether particular data points are to be included or excluded from cladistic analysis.

Modern data sets are formed from characters scored as discrete codes in columns and taxa in rows. Hence filters operate between the initial discovery procedure and the recording of the variation in a data matrix. Details of the filters used are often obscured by the style of publication and often the final reworked matrix is the only published information on the original observations. For computerized cladistic analysis, all matrices render variation into discrete codes. Frequently, only characters that show significant variation are used and measurements are coded into discrete states on the basis of gaps. These gaps may be the only clues to the type of the filter used.

2.1.1 Filters

The most obvious filter in cladistic analysis is that which rejects attributes that are continuous and quantitative and favours instead characters that are discrete and qualitative. The problem with all characters is determining those that are cladistically useful and those that are not. In general, continuous and quantitative characters are considered not to be cladistic but to vary phenetically. There are many reasons why we favour discrete characters and consider continuous characters unsuitable for cladistic analysis. Quantitative characters are difficult to describe fully, requiring means, medians and variances to establish the gaps. Using only a portion of the described character (e.g. the mean or the median) raises the question of what to do with the rest, but matrices generally require that each value in the matrix be represented by single discrete alphanumeric value, although polymorphic variables can be analysed in certain computer programs (e.g. PAUP and MacClade).

2.2 KINDS OF CHARACTERS

The recommendation to reject quantitative and continuous data in favour of qualitative and discrete data implies that 'quantitative', 'qualitative', 'continuous' and 'discrete' refer to kinds of data that differ in value, and that some quantitative variables can be determined *a priori* to have greater systematic value than others. This chapter reviews characters and coding from all four perspectives: qualitative and quantitative, discrete and continuous.

2.2.1 Qualitative and quantitative variables

The distinction between qualitative and quantitative data is considered by some to be more apparent than real. Stevens (1991) considered that 'many so-called qualitative characters are based on a quantitative base filtered through reified semantic discontinuities of... terminology'. What this means is that although matrices seem to be coded with discrete characters, the original observations were qualitative ranges that had been artificially filtered. In other words, qualitative terminology hides quantitative values. For example, descriptions of plane shapes, such as leaves ovoid, are shorthand expressions for ranges of measurements, in this case, measurements of the dimensional ratio between distance from the base to the widest point on the leaf. The distinction between qualitative and quantitative refers more to mode of expression rather than to intrinsic properties of the data. Table 2.1 provides a list from Thiele (1993) of quantitative characters from the literature that are expressed qualitatively with clear gaps between the different states.

Qualitative shorthand expressions of quantitative data are most useful when the data show widely discontinuous patterns of variation. Thus a sample of leaves distinguishable into ovate and obovate forms may be simply described in a quantitative way. If ovate leaves belong to one taxon and obovate leaves belong to another taxon, the character can be scored in the matrix for cladistic analysis, either as a 0 or 1, or any pair of adjacent integers. However, if the sample of specimens has a continuous range of variation between one

Table 2.1 Examples of qualitative variables expressed as discrete quantitative characters. (After Thiele 1993)

Male anal point reduced; distinct[1,3]
Larval antennal blade longer than flagellum; shorter than flagellum[2,3]
Cotyledons ovate-orbicular; spathulate[2,4]
Fruit ribs prominent; not prominent[1,4]
Buds widest below operculum suture; above operculum suture[2,5]
Fruit valves broader than long; approximately equal; longer than broad[2,5]
External valves of premaxilla midbody flat; markedly indent[1,6]
Posterolateral overlap of ectopterygoid with maxilla extensive; modest[2,6]
Dorsal fin anterior; posterior[2,7]
Vomerine teeth extending laterally at least as far as the medial borders
   of the internal nares; not extending as far[2,8]
Diastema between palatine and vomerines small; large[1,8]
Pollen boat-shaped; globose[2,9]
Nucellar cuticle thin; thick[1,9]
Pterygoid floor of canalis caroticus internus thin; thick[1,10]

---

[1] Length measurement.
[2] Ratio.
[3] Cranston and Humphries' (1988) recoding of Saether (1976).
[4] Thiele and Ladiges (1988).
[5] Chappill (1989).
[6] Kluge (1989).
[7] Begle (1991).
[8] Kraus (1988).
[9] Laconte and Stevenson (1991).
[10] Gaffney *et al.* (1991).

form and another, then the variation would require some quantitative expression to do it justice. If two taxa each have a range of leaf shapes along this continuum and the ranges overlap, the character would more often than not be labelled quantitative and rejected from cladistic analysis. The terms quantitative and qualitative are often used in this sense as synonyms for overlapping and non-overlapping ranges in variables.

2.2.2 Discrete and continuous variables

The terms discrete and continuous properly refer to mathematical properties of the range of numbers used to measure an attribute. Continuous data are those, such as dimensions, where potential values are so infinitesimally close that there are no disallowable real numbers. In contrast, discrete data can be represented logically only by a subset of all values, generally integers. They include absence/presence data (0/1 being the only allowable values), multi-state data $(0/1/2/\ldots n)$, meristic data (counts of structures expressed as

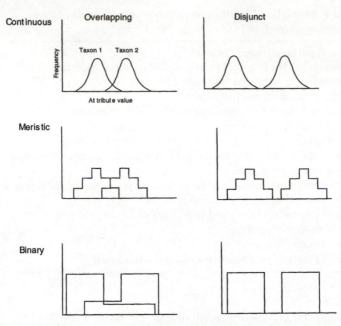

Fig. 2.1 Overlapping and non-overlapping patterns of variation in continuous, meristic and binary data. (After Thiele 1993.)

integers, directly scored into the matrix or rescaled) and molecular data (e.g. nucleotide sequences, ACGT/U).

2.2.3 Overlapping and non-overlapping characters

While qualitative, quantitative, discrete and continuous are useful terms, the degree of overlap among them is the crucial property (Fig. 2.1). Although it is implied that overlap can occur only between continuous characters, both continuous and discrete characters can exhibit different degrees of overlap. It is the degree of overlap that makes the distinction in filtering between overlapping and non-overlapping characters. The filtering proscription can be set to particular values for any given character. For example, it can be set to select only those characters that show no overlap, rejecting all others. The problem remains, however, because, in reality, we have a sliding scale from widely overlapping characters to widely disjunct characters that have discrete gaps between them. The required filter in these situations is a cut-off point where the critical value might be scored for no overlap or any other arbitrarily selected value. The problem is that this makes the filtering of characters highly susceptible to sampling error and the cut-off points between characters quite arbitrary. In reality, we should be able to grade characters along a sliding scale and develop methods of coping with different degrees of overlap. The sliding scale could recognize non-overlapping data as better than overlapping data and that the latter should be used only when the former are unavailable. This approach matches the continuum of degree of overlap with a continuum from better to worse (Chappill 1989), rather than forcing it into the general division of good and bad characters (Pimentel and Riggins 1987; Thiele 1993).

2.3 CLADISTIC CHARACTERS

The concept of a character is ill defined in cladistics and has a multiplicity of meanings. Viewing features of organisms as characters or character states is part of the process of recognizing their systematic value when distinguishing them from non-cladistic characters. Characters are generally listed in the form of presences or absences (e.g. vertebrae present/vertebrae absent), as binary variables expressed as alternative characters or character states (e.g. anthers introrse/anthers extrorse), or as multistates (e.g. eyes blue/eyes green/eyes brown). Although these distinctions might be obvious for any one group of organisms, it is impossible to write down how a character or character state might be defined in terms of its systematic value. For, inasmuch as we discover organisms and develop hypotheses of the boundaries and relationships of taxa through the study of collected samples in both living and preserved collections, so the same applies to characters. Initial hypotheses of primary homology are subjected to similar processes to become refined into secondary homologies (synapomorphies) through successive cladistic analyses and modifications of the original hypotheses (de Pinna 1991). Seen in this way, characters are hypotheses about structures or features that can be put through cladistic analysis to determine whether they are homologues or not.

2.3.1 Diagnostic and systematic characters

There has always been a tension between the notion of defining characters in order to identify and distinguish organisms and the discovery of homologies in comparative biology to systematize the relations among organisms (Table 2.2). Thus Smith (1994*a*), for example, considered that characters must occur in two or more states (one of which may be absence) and should be defined as objectively as possible. Both morphological and molecular features that are indistinguishable are generally coded as the same character state so as to reflect the underlying notion of primary homology. The problem with such definitions (Table 2.2a) is that they do not distinguish between characters used in keys for the express purpose of distinguishing a taxon from any of its relatives and those characters that are homologues and suggest relationships.

Table 2.2 Definitions of characters

a. Characters diagnostic of taxa

'Characters are observed variations which provide diagnostic features for differentiation amongst taxa' (Smith 1994 a).
A character is 'any attribute of an organism or a group of organisms by which it differs from an organism belonging to a different category or resembles an organism of the same category' (Mayr *et al.* 1953).
A character is 'anything that is considered a variable independent of any other thing considered at the same time' (Cain and Harrison 1958).
'A character in systematics may be defined as any feature which may be used to distinguish one taxon from another' (Mayr *et al.* 1953).
A character is 'a feature of an organism that is divisible into at least two conditions (or states) and that is used for constructing classifications and associated activities (principally identification)' (Stuessy 1990).

b. Characters as transformations

'A character is a feature of an organism which is the product of an ontogenetic or cytogenetic sequence of previously existing features, or a feature of a previously existing parental organism(s). Such features arise in evolution by the modification of previously existing ontogenetic or cytogenetic or molecular sequence' (Wiley 1981).
'A character is a feature of an organism that can be evaluated as a variable with two or more mutually exclusive and ordered states' (Pimentel and Riggins 1987).
'A character ('transformation series' of Hennig) is a collection of mutually exclusive states (attributes; features; 'characters', 'character states', or 'stages of expression' of Hennig) which
a) have a fixed order of evolution such that
b) each state is derived directly from just one other state, and
c) there is a unique state from which every other is ultimately derived'
(Farris *et al.* 1970).

c. Attributes of organisms, character states of taxa

'Attribute states are 'the descriptive terms which are applied to individual organisms, e.g. 'red', '2 cm long''; attributes are 'sets of such descriptive terms, e.g. 'colour' to which 'red' and 'green' belong'; character states are 'probability distributions over the states of an attribute'; characters are 'sets of such probability distributions'' (Jardine 1969).

d. Characters as homologues

A character is 'a theory that two attributes which appear different in some way are nevertheless the same (or homologous)' (Platnick 1979).
'If... characters are hypotheses of homology and synapomorphy, then they must be relational, and the units of these relations are three-taxon statements' (Nelson and Platnick 1991).
'Cladistics is a discovery procedure, and its discoveries are characters (homologies) and taxa' (Nelson and Patterson 1993).

2.3.2 Character transformations

Nevertheless, for characters or character states to be cladistic, and hence be features of taxa, they must be scorable in a data matrix and contain some pattern for hypotheses of relationships of taxa to be discovered. For evolutionary biologists, characters transform from one condition into another. For example, Wiley's (1981) definition recognized that features of organisms are the products of evolution and hence have arisen as changes in ontogeny and transformation through time. However, there is a problem because this definition is one of transformation of one character or character state into another within organisms, rather than of an homology of any particular group. Thus Wiley used his definition to describe characters of Chordata and Vertebrata, which are clearly taxa consisting of many individual organisms, rather than transformations of features within organisms.

Pimentel and Riggins (1987) were stricter but less rigorous in their definition when they stated that a character can only be a feature of an organism when it can be recognized as a distinct variable. Their definition is also problematic because it, too, is tied to features of organisms rather than taxa and, like Wiley, they go on to discuss coding variables for taxa rather than for organisms. Farris (quoted in Mitter 1980) showed that determining characters was an inductive process when he stated that 'morphologists do not sample characters, they synthesize them'. The Pimentel and Riggins definition is based on that of Farris *et al.* (1970), who made it clear that, in order to be able to determine characters for phylogenetic reconstruction, it was necessary to recognize that they were mutually exclusive states that could be considered transformations with a fixed order of evolution. Farris *et al.* thus redefined Hennig's Darwinian interpretation, that characters transform from one state to another, as a series of axioms. All of the definitions in Table 2.2a–b confuse the relationship between organisms and taxa and the problem remains as to what diagnoses taxa when definitions refer to attributes of organisms.

2.3.3 Characters and character states

Jardine (1969) considered that diagnosing taxa and individual organisms using the same character was nonsensical. The presence of a backbone is not a property of Vertebrata but all of the organisms within the group Vertebrata possess backbones. Jardine made the distinction between taxa and organisms by describing characters and character states. Taxa have characters and organisms have attributes or character states (Table 2.2c).

Most cladists consider that for characters to convey cladistic information they must transform from one state into another through time. However, this does not mean that a green eye changes into a blue eye or that oval leaves change into obovate leaves. Similarly, invertebrate animals lacking backbones do not change into those that possess them. What actually changes is the frequency of a particular character state for a given character and the frequencies of different character states change through time. Cladistic character states are frequency distributions and, conversely, all cladistic character states have particular frequencies of distribution. Thus, desirable cladistic characters are those with large, clear-cut changes rather than small, gradual ones, and a good cladistic character is, in effect, a value judgement on data.

2.3.4 Homology

The desire for cladistic characters to express large, clear-cut differences between taxa does not go far enough in determining which character states become grouping homologies. To an evolutionist, homology is defined as the same structure inherited from a common ancestor. Thus to Hennig, hypotheses about characters (synapomorphy) and hypotheses about groups (monophyly) both appealed to ancestors for their justification. This concept was shown to lead to circular reasoning because both hypotheses for characters and hypotheses for groups appeal to the same mysterious non-empirical ancestors. The solution came with the so-called 'transformation of cladistics', which allowed hypotheses about character states (homology) to give hypotheses about groups (hierarchy). The method is empirical in that there are no appeals to ancestry for the determination of monophyletic groups (Platnick 1979) and any interpretations about ancestry are derived from the cladogram.

For cladistic analysis to be successful, we consider that it is necessary not only to have principles that do not assume transformation, but also to describe characters as hypotheses of homology that can be tested (Table 2.2d). Homology is the core concept of comparative biology and systematics. When comparing and contrasting the morphology and anatomy of organisms, we break down our observations into traits or character states as recognizable features of the whole organism. Characters and character states convey no phylogenetic information until we recognize their existence in other organisms through naming them (Patterson 1982). It is the act of naming characters and character states that establishes theories of homology.

An hypothesis of homology recognizes that a character in one taxon represents the 'same' feature as a similar, but often not identical, character in another taxon. Structures that are identical in form, position and development in two or more organisms pose no problem, because different systematists can agree that they represent the same entities, i.e. they have a clear one-to-one correspondence. However, problems arise when structures have diverged in form so as to be only vaguely similar or when different developmental pathways arrive at similar structures. Proposing hypotheses of homology becomes critical when we come to define 'sameness' among structures. Similarity of form is not a criterion of homology in itself but a first-order hypothesis. For an homology to become an established character the feature

Table 2.3 Tests of homology

Test	Relation		
	Homology	Parallelism	Convergence
Similarity	+	+	—
Conjunction	+	+	+
Congruence	+	—	—

in question must also occur in the same topographical position within the organisms being compared and also agree with other characters about relationships of taxa (character congruence), a test that can be applied only after, or during, cladistic analyses (Table 2.3).

The congruence test equates homology with synapomorphy. Characters that fit to a cladogram with the same length pass the test, whereas those requiring more steps are deemed homoplastic. Thus, determination of homologues becomes an empirical procedure and the final arbiters of homology are the characters and character states themselves (Patterson 1982). Consequently, the more characters that are included in the analysis, the more demanding the test of homology becomes. This aspect becomes important in later considerations of simultaneous analysis or so-called total evidence (see Chapter 6).

2.4 CHARACTER CODING FOR DISCRETE VARIABLES

The crucial point in systematics is not the source of evidence from which we attempt to derive characters, as this will vary from group to group, but how features might be usefully coded so as to reflect accurately our observations for a particular scale of problem. We have already stated that for characters or character states to be cladistic, and hence features of taxa that might be potential homologues, they must be scorable into data matrices that contain some pattern for relationships of taxa to be discovered. Converting raw data into codes for cladistic analysis is something that has to be done with considerable care. There are many different ways of coding characters and the outcome of different coding schemes can dramatically affect hypotheses of relationships. We shall demonstrate this using a hypothetical example of conflicting character states of shape (round and square) and colour (black or white) (Fig. 2.2).

For the purposes of exposing some of the problems associated with character coding, we describe the following four coding methods, although they do not exhaust the possibilities:

a linked multistate character (Table 2.4a);

two separate multistate characters for shape and colour (Table 2.4b);

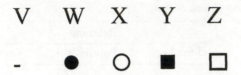

Fig. 2.2 Different expressions of conflicting characters in five taxa (V–Z): absent, round and black, round and white, square and black, square and white. (After Pleijel 1995.)

three independent binary characters for shape and colour (Table 2.4c);

five completely independent characters (absence/presence coding) (Table 2.4d).

Table 2.5 shows how these characters and character states might be distributed among five different taxa (V–Z). Coding method A assumes interdependence between the main features and codes everything into a single multistate character. Coding method B treats colour and shape as two quite separate characters but includes an extra state (0) to account for absence of each character. Coding method C is similar to B but treats

Table 2.4 Four coding methods for the features shown in Fig. 2.2. Characters are labelled with integers in bold and character state codes as integers in parentheses

---

(a) *Method A: formula coding as one multistate character with linked states.*

1. absent (0); round and black (1); round and white (2); square and black (3); square and white (4)

(b) *Method B: shape and colour attributes treated as two independent multistate characters.*

1. absent (0); round (1); square (2)
2. absent (0); black (1); white (2)

(c) *Method C: hierarchical coding with shape and colour attributes treated as two independent binary characters and an additional code for presence and absence of the features; inapplicable observations for the absence of the feature accommodated using question marks.*

1. features absent (0); features present (1)
2. round (0); square (1)
3. black (0); white (1)

(d) *Method D: independent coding of variables assuming no transformations.*

1. features absent (0); features present (1)
2. round shape absent (0); round shape present (1)
3. square shape absent (0); square shape present (1)
4. black pigmentation absent (0); black pigmentation present (1)
5. white pigmentation absent (0); white pigmentation present (1)

Table 2.5 Five taxa (V–Z) with the features shown in Fig. 2.2, scored according to the four coding methods (A–D) listed in Table 2.4

Taxa	Coding method			
	A	B	C	D
V	0	00	0??	00000
W	1	11	100	11010
X	2	12	101	11001
Y	3	21	110	10110
Z	4	22	111	10101

presence and absence of any character state as a separate character and thus has three columns. Coding method D assumes that all five character 'states' are independent characters. Following Pleijel (1995), these coding methods are discussed under the four different headings of character linkage (dependency between characters in a single matrix), hierarchical dependency, missing values and information content.

2.4.1 Multistate characters—character linkage during analysis

It is generally held that characters in a cladistic analysis should represent independent hypotheses of relationship. Correctly identified homologies are expected to exhibit congruence and converge onto similar fundamental cladograms (ideally one). Homoplastic characters, on the other hand, are expected to show scatter on cladograms. The main problem in choosing an appropriate coding method is to arrive at an accurate division of characters and character states so that they reflect the relationships of the organisms. The more that characters become linked the greater is the departure from independence and consequently the risk that one false homology can obscure the topologies of true homologies is decreased. Uniting independent features into a single multistate character (method A) minimizes the effects of linkage, while treating multistate characters as unordered (see Chapter 3) allows transformation of one character state to another in a single step. Because there are four states in the example (in addition to absence), there are four possible types of transformations: a single ordered multistate, a single unordered multistate, two ordered multistates, and two unordered multistates.

2.4.2 Binary characters—character linkage during analysis

In all other methods, where the data are partitioned into separate columns, the degree of partition creates different problems. Method B treats shape and colour as independent characters but to cater for those taxa that do not have any of the features, it is necessary to code for their absence using an extra state (0) (Table 2.5). A major effect of this approach is that duplication of absences might become a problem when many different characters are perceived as connected to a feature that is absent from some taxa (Maddison 1993). This problem is solved to some extent by method C, where presence/absence is included as a separate character and missing values are used for inapplicable observations (Table 2.5). Coding method D makes no adjustment for character linkage and treats the characters as five separate columns in the matrix (Table 2.5).

2.4.3 Hierarchical character linkage

Characters and character states will often be coded differently in an analysis of a group of closely related taxa when compared to a more general study that contains these taxa as only one small subgroup. At the more general level, we may be satisfied to code just that information relating to the absence or presence of a feature, whereas at the less general level we might choose to encode more of the observed variation. Thus, characters and character states can vary in how they are coded at different scales, in other words, character interdependency is affected by hierarchy.

Coding method A divides the features of shape and colour into five linked states in a single character. This may be satisfactory for a high level analysis, when the main distinction is between absence and presence of any of the different states. However, at a more inclusive level, we would tend to code the shape and colour as separate characters and states, especially if none of members of the study group lack any of the states being considered. The four coding methods, A–D, can be placed on a sliding scale of differential reformulation to satisfy different analytical scales. The addition and deletion of character states at different levels of analysis will affect the cladogram topologies found using methods A–C much more than with method D, which, by dividing the component attributes as finely as possible, tends to remain stable at any level of analysis.

2.4.4 Transformation between character states: order and polarity

Once characters are coded as multistates, the implications of order and polarity can have major effects. Consider a character with three states, 0, 1 and 2 (Fig. 2.3). In ordered analysis, the gain or loss of a state may be viewed as incremental and thus a change from 0 to 1, or from 1 to 2, is considered as one step. A change from state 0 to state 2 requires a transformation via state 1 and requires two steps. The character is said to be ordered: $0 \leftrightarrow 1 \leftrightarrow 2$. If the direction of change of such a character is also fixed using an *a priori* criterion, then the character is also said to be polarized, in this example either as $0 \rightarrow 1 \rightarrow 2$, $2 \rightarrow 1 \rightarrow 0$ or $1 \leftarrow 0 \rightarrow 2$. In contrast, unordered characters are allowed to change from any one state into any other state with equal cost and thus nine transformations are possible (Fig. 2.3). These possibilities a   b   c

$0 \to 1 \to 2$ $0 \to 1 \to 2$ $0 \to 1 \to 2$

$0 \to 2 \to 1$ $2 \to 1 \to 0$

$1 \to 0 \to 2$ $1 \leftarrow 0 \to 2$

$1 \to 2 \to 0$

$2 \to 1 \to 0$

$2 \to 0 \to 1$

$0 \leftarrow 1 \to 2$

$0 \leftarrow 2 \to 1$

$1 \leftarrow 0 \to 2$

Fig. 2.3 (a) The nine possible transformations for a multistate character with three states: 0, 1, 2. (b) The three allowable transformations between three states following imposition of the order shown in (c).

can be restricted to a set of three transformations by selecting a particular character order and then to a single sequence by making another choice regarding polarity. (See Chapter 3 for a detailed account of polarity, rooting and optimization).

For example, consider that through study of the ontogeny, we have ascertained that not only is absence plesiomorphic but also that the round shape is subsequently transformed into the square shape. The states of this multistate character could now be considered incremental and both the order and polarity could be included as extra information in a cladistic analysis.

2.4.5 Missing values and coding

The problem with linked characters due to absences may be circumvented by adding extra absent/present characters, as in Table 2.4c. Taxa lacking the features of both shape and colour are then scored with question marks (i.e. as missing values) for character states connected to this feature (Table 2.5). This solution draws attention to other problems. Platnick *et al.* (1991a) showed that absences can occur for a variety of reasons: unknown data, inapplicable data and polymorphism (see also §4.2). Coding problems due to terminal polymorphism can be catered for in programs such as PAUP and MacClade, while the problem of unknown data can only be solved by further observation. However, the dilemma of inapplicable data remains (Pleijel 1995). Coding methods A and B simply accommodate the problem by treating absence as a state equivalent to both shape and colour, while in method D, the problem cannot exist. In method C, the use of question marks does not distinguish between inapplicability and absence due to lack of knowledge and, furthermore, can lead to problems in interpretation of results (Platnick *et al.* 1991a). A detailed account of the problems of missing data is given in Chapter 4.

Table 2.6 Sankoff cost matrix (see text for information). Character codes follow Table 2.4; absent (0); round and black (1); round and white (2); square and black (3); square and white (4). Character costs are shown as 0, 1 or 2 in the matrix

	0	1	2	3	4
0	0	2	2	2	2
1	1	0	1	1	2
2	1	1	0	2	1
3	1	1	2	0	1
4	1	2	1	1	0

2.4.6 Information content and the congruence test

For any coding method to be efficient, it must be able to transform observations into a suitable form for cladistic analysis without loss of information, so that the coded information can take part in the most critical test of homology: congruence. The single multistate coding procedure (method A) (Table 2.5a) intimately ties colour and shape into a formula. Thus, it can never allow either of the features to be tested independently of one another by other characters. However, because there are four states, two for colour and two for shape, it is possible to code a cost matrix (Sankoff and Rousseau 1975) that considers the logical transformations (Table 2.6). A change from absence to presence of any shape and colour costs two steps. However, change from one shape to another, or from one colour to another, costs one step. Loss of the feature also costs one step. Such cost matrices make no assumptions about the direction of transfomations, but allow differential costs to be applied to different state changes within a single multistate character. Cost matrices are discussed further under the heading of 'generalized optimization' in §4.1.5.

In contrast, absence/presence coding (method D) (Table 2.5d) is a formulation of all potential homologies. Methods B and C (Tables 2.4b–c) are problematic in that they make transformation assumptions in the codings that cannot be tested by congruence. Character states represent homologies in the formulation of a character that are tested by congruence of similar character state codings of other characters. However, if two character states are locked into the same character, e.g. round with square or black with white, at least one of the alternatives can never be tested by congruence. Instead, the homology of the two states is assumed *a priori*. The absence/presence coding method avoids making homology assumptions between features by coding them all as separate characters. This method of coding allows every character to be tested by every other character, but ignores the logical dependency between the two manifestations of shape and the two of colour. It denies the primary homology assessment based on similarity and can lead to pseudo-parsimonious reconstructions during cladistic analysis (Meier, 1994).

2.5 MORPHOMETRIC DATA IN CLADISTIC ANALYSIS

All character states (as used in cladistic analysis) are frequency distributions of attribute values over a sample of individuals of a taxon (Thiele 1993). Consequently, there are many situations in which continuously variable morphometric data have to be considered as cladistic characters, even when the taxa have overlapping frequency distributions. It is quite possible that continuous values, however opaque in the raw form, contain grouping homologies when identified through discrete coding and cladistic analysis. Continuous variables should only be excluded if the cladistic analysis cannot handle such data or if it can be shown empirically that those characters convey no information or phylogenetic signal relative to other characters in the data matrix.

2.5.1 Coding morphometric data

There are a number of methods (e.g. in MacClade) that can handle morphometric data without recoding and that have some limited use for looking at character evolution over trees (e.g. Swofford and Berlocher 1987, Huey and Bennett 1987). However, in order to compare continuous variables with qualitative variables on cladograms, continuous characters have often been recoded as discrete characters (e.g. Cranston and Humphries 1988, Chappill 1989, Thiele and Ladiges 1988).

Generally, all methods can be described as gap-coding methods, although there are numerous variations: simple gap-coding (Mickevich and Johnson 1976); segment coding (Colless 1980); divergence coding (Thorpe 1984, Almeida and Bisby 1984); generalized gap-coding (Archie 1985, Goldman 1988); range coding (Baum 1988) and gap-weighting (Thiele 1993). They all have one thing in common—a simple algorithm to create gaps so as to produce discrete codes for overlapping or continuous values. Samples of taxa are ranked along a scaled attribute axis, and then the attribute axis is divided into states. Simple gap-coding divides the axis at those points where no values occur or between the means of the frequency distributions at the point where the 'gap' exceeds a particular preconceived value, such as one standard deviation about the mean. Usually, the attribute axis will be divided into fewer states than there are taxa and for most computer programs there is an upper bound to the number of states per character that can be analysed (32 in PAUP, 26 in MacClade and 10 in Hennig86, PIWE and NONA).

2.5.2 Gap-weighting

The following method from Thiele (1993) is one of several methods that provides more than simple gap-coding by adding a weight code. Gap-weighting uses additive coding to add steps to each code so that the score in

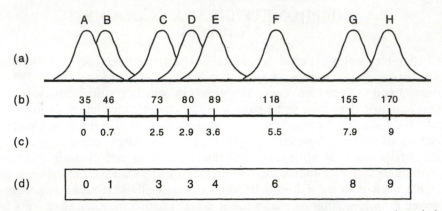

Fig. 2.4 Example of coding using the gap weighting method of Thiele (1993). (a) Frequency distribution curves for eight taxa, A–H. (b) Means for the taxa on the attribute scale. (c) Values scaled to a range of 10 (0–9). (d) Integer coding for analysis using Hennig86.

the column of the data matrix not only relates to the position of each state relative to every other state over the range, but also maintains the relative sizes of the gaps between them. A suitable rescaling function is also used to allow the full range of integers that can be handled by a given cladistic computer programme to be used and thus ensure that as much of the raw attribute data as possible is utilized in the codes.

1. The raw data are initially ranked as an ordered set of states, arranged according to the values of the means, medians or other appropriate measure of range. If the variances are not equal, the data should be standardized using an appropriate transformation. One of the simplest is $\log(x + 1)$ (Fig. 2.4a).

2. The data are then range standardized, e.g.:

$$x_s = (x - \text{min}/\text{max} - \text{min})n$$

where x is the raw datum, x_s is the standardized datum and n is the maximum number of ordered states allowed by the cladistic computer program (Fig. 2.4b).

3. Code the values as the rounded integer of the standardized values (Fig. 2.4c, d).

4. Treat the character as an ordered multistate for analysis (Fig. 2.4d).

Thiele (1993) considered the essential elements of the method to be as follows.

1. The rescored characters retain information on both the rank order of

states and the sizes of the gaps between states. Consequently, transformations between states are weighted proportionally to the sizes of the gaps separating them.

2. All differences between states are accepted as potentially informative. Parsimony is relied upon to discriminate truly informative gaps from spurious ones, rather than using *a priori* statistical tests.

3. Differences within and between characters are equalized using a transformation and range-standardization procedure.

The method weights explicitly, using the assumption that transformations between widely separated states are more likely to be informative than those between narrowly separated states. This is important when characters conflict, especially different continuous variables. However, this assumption will be contested when broader patterns of congruence become the final arbiter during analysis. In a conventional analysis, binary 0/1 characters have a range of 0 to 1. Gap coded characters have ranges of 0–9, 0–26 or 0–32 depending upon the computer program used. It is as important, therefore, that binary characters are either weighted by 10, 26 or 32 or coded as 0/9, 0/26 or 0/31, so as to maintain parity.

2.6 DISCUSSION: CHARACTER DISCOVERY AND CODING

The aims of character discovery and character coding are to identify as accurately as possible those features that ultimately diagnose relationships of taxa. Characters should be determined and scored so that all possible hypotheses of homology or synapomorphy can be examined through cladistic analysis. Characters come from many sources and the aims of comparative biology and systematics are one and the same—to determine the relevant level of universality at which particular characters should be placed on cladograms so as to provide hypotheses of relationships between organisms and groups.

2.6.1 Choice of characters

The debate as to what is a good cladistic character has produced a scale of preferences from clear-cut qualitative differences that prove to be robust homologies to quantitative variables that need to be heavily manipulated using special coding procedures to extract a potential phylogenetic signal. In almost all reported analyses, it is the clear-cut qualitative characters that delimit groups unambiguously. No doubt it is true that qualitative data are more reliable than others and perhaps the perfect cladistic character is one that has an exact fit to the shortest cladogram. Continuous characters invariably produce cladograms with lower levels of fit than qualitative characters. Cladists wishing to use continuous characters have employed various procedures in order to include them. There is often significant covariation with other characters, even when continuous characters are recoded as ordered multistates, which suggests that they do tend to operate as linear series. In many cases, morphometric and qualitative characters are found to map similar phylogenies and be informative about those phylogenies. It is likely that morphometric data will continue to be used most often in studies of closely related taxa, while presence/absence characters will be used in studies of higher ranking taxa. The judgement that all morphometric data is garbage is unnecessarily harsh and it is still open to debate what constitutes reliable evidence in cladistic analysis.

2.6.2 Coding

There are two schools of thought on coding methods: those that advocate absence/presence coding and those that consider additivity and multistate coding as appropriate for diagnosis of taxic relations. Absence/presence coding is invariably binary and contrasts presence against absence. There are very few cases of true absence/presence coding in the literature, most likely due to linkage problems (Pleijel 1995). The most commonly used forms of coding are methods A, B and C (Table 2.4). Pimentel and Riggins (1987) considered that all cladistic characters should be treated as multistates and ideally coded as multiple column additive binary characters in order to distinguish *a priori* between linear and branched character state trees. They considered that character states cannot be treated as simple, nominal variables because redundancy is introduced and information content is sacrificed. In other words, additivity is a form of information and there are many reasons to be sympathetic to this viewpoint, especially in special cases, such nucleotide sequences. Here, the nucleotide codes (A, C, G, T) are invariably considered as alternatives in multistate columns. Application of absence/presence coding has yet to be considered in molecular systematics and there is no body of opinion that considers base substitution as anything other than a special form of character state transformation.

The issue of whether one uses multistate or binary coding revolves around the issue of transformation between character states (Wilkinson 1995). Devotees of multistate coding accept that characters should be treated as transformation series and that hypotheses of adjacency between similar, but different, character states should be coded *a priori* in the character state matrix. Transformation series analysis (TSA) is perhaps the most elaborate manifestation of this method (Mickevich 1982). In contrast, absence/presence coding is a more simple and straightforward approach than any of the alternatives. Every variable is kept separate as a potential synapomorphy to be tested against other coded observations. Although redundancy of additional absence scores may be a problem, there are advantages in not building unwarranted assumptions into the data. The advantage is that instead of making decisions *a priori*, character hierarchies emerge from the results (Pleijel 1995). Three-item statements analysis (Chapter 7) takes the argument further and by converting characters into minimal expressions of the kind A(BC), aims to move away from ideas of transformation in cladistic analysis.

2.7 CHAPTER SUMMARY

1. Cladistics is the discovery or selection of characters and taxa, coding of characters and determination of cladograms using the property of homology and the criterion of parsimony to best explain the distribution of characters over the taxa.

2. Modern data sets generally require characters to be scored as discrete alphanumeric codes in columns and taxa in rows.

3. A variety of filters operate between the initial discovery procedure and the recording of variation in a data matrix. The most commonly used filter in cladistic analysis tends to reject characters that are continuous and quantitative and favours instead characters that are discrete and qualitative. However, it is demonstrated here that the main issue is to determine for all characters which are cladistically useful or not, and that all characters can be arranged on a sliding scale of most useful to least useful. To reject quantitative and continuous data in favour of qualitative and discrete data implies that *quantitative, qualitative, continuous* and *discrete* refer to different kinds of data differing in value, and that some quantitative variables can be determined *a priori* to have greater systematic value than others.

4. The interplay between the discovery procedure of determining characters and cladistic analysis determines which structures are homologues and which are not. Homology is the core concept of comparative biology and systematics. For cladistic analysis to be successful we believe that it is necessary to have principles that do not assume transformation, but to describe characters as hypotheses of homology that can be tested with the criteria of similarity, conjunction and congruence.

5. There are many different ways of coding characters and the outcomes of different coding schemes can dramatically affect hypotheses of relationships. For discrete characters four coding methods are described to show the differences between various kinds of multistate coding and binary coding. Character coding methods are discussed under four different aspects of character linkage (or dependency between characters in a single matrix), hierarchical dependency, missing values and information content, to show that coding methods that offer the most stringent test of homology are those to be preferred. Finally, there is a brief discussion of the principles of coding continuous characters.

3.
Cladogram construction, character polarity and rooting

3.1 DISCOVERING THE MOST PARSIMONIOUS CLADOGRAMS

3.1.1 Hennigian argumentation

The first explicit method of cladogram construction was proposed by Hennig (1950, 1966) and is thus called Hennigian argumentation. This method considers the information provided by each character individually. Groups on cladograms are recognized by the possession of apomorphies and, in Hennigian argumentation, these apomorphies are identified *a priori*, that is, the characters are polarized into plesiomorphic and apomorphic states before the cladogram is constructed. The subject of character polarity and the recognition of apomorphies is discussed in detail later in this chapter. For now, it is assumed that the apomorphies have been recognized and coded 1, with the plesiomorphic states coded 0.

Consider the data set in Table 3.1 (derived from the characters in Fig. 1.3, p. 4), in which we wish to resolve the interrelationships of the shark, salmon and lizard. This group comprises the study taxa, also referred to as the ingroup. Taxa with which members of the ingroup are compared in order to polarize characters are termed outgroup taxa, in this case, the lamprey. First, consider character 1. All the ingroup taxa share the apomorphic state, in contrast to the outgroup, which has the plesiomorphic state. Character 1 thus unites the ingroup as a monophyletic group (Fig. 3.1a). Character 2 shows the same distribution and corroborates the information contained in character 1. Repeating this approach, characters 3 and 4 can be seen to unite the salmon and the lizard into a monophyletic group to the exclusion of both the lamprey and the shark (Fig. 3.1b). The relationships among the ingroup taxa are now fully resolved. Characters 5–12 are autapomorphies of individual ingroup taxa and thus have no role in cladogram construction at this level of universality (Fig. 3.1c). However, if this solution is accepted, then character 13 must be placed on to the cladogram twice, resulting in a total of 14 steps. An alternative topology would unite the shark and the salmon using the putative apomorphic state of character 13 (Fig. 3.1d). However, as seen in Chapter 1, this topology would entail two occurrences each for characters 3 and 4. It would therefore have 15 steps and would be rejected as a less parsimonious result.

Table 3.1 Data set to illustrate Hennigian argumentation

Lamprey	0000000000010
Shark	1100000000101
Salmon	1111000001001
Lizard	1111111110000

With small data sets that are reasonably free from homoplasy, Hennigian argumentation is quick and simple to implement. However, most data sets have large numbers of taxa and characters, as well as greater degrees of homoplasy, which makes finding the most parsimonious cladograms by Hennigian argumentation extremely time-consuming. Thus, computerized methods have been developed that speed up the search for most parsimonious or minimum-length cladograms.

3.1.2 Exact methods

Such computerized methods fall into two categories. Exact methods are those that guarantee to find one or all of the shortest cladograms. Of these, the simplest to understand is exhaustive search, in which every possible fully resolved, unrooted cladogram for all the included taxa is examined and its length calculated. In this way, it is certain that all minimum-length cladograms will be identified. A simple algorithm to perform an exhaustive search

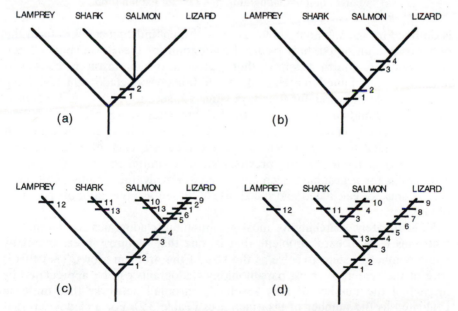

Fig. 3.1 (a–d) Determination of the most parsimonious cladogram for the data in Table 3.1 using Hennigian argumentation. See text for explanation.

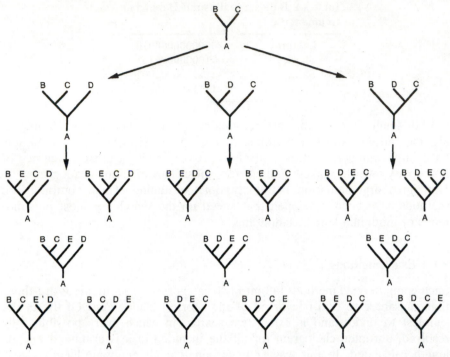

Fig. 3.2 Illustration of the exhaustive search strategy for determination of most parsimonious cladograms. See text for explanation.

is outlined in Fig. 3.2. First, three taxa are chosen and connected to form the only possible unrooted, fully resolved cladogram for these taxa (Fig. 3.2, top cladogram). A fourth taxon is then selected (which taxon is chosen is immaterial) and joined to each of the three branches of cladogram 1, yielding three possible networks for four taxa (Fig. 3.2, second row). A fifth taxon is then selected and added to each of the five branches of the three cladograms, giving 15 cladograms (Fig. 3.2, rows 3–5). This procedure, in which the nth taxon is added to every branch of every cladogram (each of which contain $n-1$ taxa) generated in the previous step, is continued until all possible cladograms for n taxa have been constructed. Finally, the lengths of all these cladograms are calculated and the shortest chosen as optimal (most parsimonious).

Unfortunately, searching for most parsimonious cladograms is what mathematicians term a 'hard' problem, that is, one that requires an exponentially rising number of steps to solve as the size of the problem grows. That this is true of the search for most parsimonious cladograms can be appreciated by inspecting the number of fully resolved, unrooted networks that must be evaluated as the number of taxa increases (Table 3.2). For a cladogram that currently includes $n-1$ taxa, there are $2n-5$ possible positions to which the nth taxon can be attached. So, while there are 105 fully resolved, unrooted

Table 3.2 The number of fully resolved, unrooted networks possible for n taxa

n	
1	—
2	1
3	1
4	3
5	15
6	105
7	945
8	10 395
9	135 135
10	2 027 025
11	34 459 425
12	654 729 075
13	13 749 310 575
14	316 234 143 225
15	7 905 853 580 625
16	213 458 046 676 875
17	6 190 283 353 629 375
18	191 898 783 962 510 625
19	6 332 659 870 762 850 625
20	221 643 095 476 699 771 875
62	$6.664\,094\,61 \times 10^{98}$
63	$> 10^{100}$

cladograms for six taxa, there are over 2×10^{20} topologies for only 20 taxa, while the number exceeds 10^{100} for as few as 63 taxa. Thus, it is doubtful whether exhaustive search is a practical option for problems with more than a moderate number of taxa.

Fortunately, there is an exact method available that does not require every completed topology to be examined individually—the branch-and-bound method. The branch-and-bound procedure closely resembles that of exhaustive search, but begins with the calculation of a cladogram using one of the heuristic methods described later in this chapter. The length of this cladogram is retained as a reference length or upper bound for use during subsequent cladogram construction. The branch-and-bound method then proceeds in a similar manner to exhaustive search but now, as the path is followed, the lengths of the partial networks are calculated at each step and compared with that of the upper bound. As soon as the length of a partial network exceeds that of the upper bound, that path of cladogram construction is abandoned, because the attachment of additional taxa can serve only to increase the length further. By this means, the number of completed cladograms that must be evaluated is greatly reduced.

Once a particular path is completed and all taxa have been added, then the length of the resultant cladogram is once more compared with the upper bound. If its length is equal to the upper bound, then this cladogram is retained as one of the set of optimal topologies and the branch-and-bound process continued. However, if the length is less than the upper bound, then this topology is an improvement and its length is substituted as the new upper bound. This substitution procedure is important because it enables subsequent paths to be abandoned more quickly. Once all possible paths have been examined, then the set of optimal cladograms will have been found.

It is impossible *a priori* to estimate the exact time required to undertake a branch-and-bound analysis, as this is a complex function of computer processor speed, algorithmic efficiency and the structure of any homoplasy in the data. Most branch-and-bound applications employ algorithmic devices to ensure the early abandonment of path searches and thus reduce computation time. For example, efficient heuristic methods can minimize the initial estimate of the upper bound. However, a branch-and-bound analysis is still time consuming to implement and should not generally be considered for data sets comprising large numbers of taxa.

3.1.3 Heuristic methods

For larger data sets, approximate or heuristic methods must be adopted, which generally use 'hill-climbing' techniques. These approaches are essentially trial-and-error and do not guarantee to find all, or even any, of the minimum-length cladograms. Thus certainty of finding the optimal result is sacrificed in favour of reduced computational time.

As an analogy, consider a group of hikers who are aiming to climb to the top of a mountain as fast as possible, with the aid of a map and a compass. However, as they begin their ascent, they walk into a mist that obscures their view of all but the immediate vicinity. In order to reach the top in the shortest possible time, the hikers' best strategy would be to walk up the mountain following the line of steepest ascent. The map shows them that this line is always perpendicular to the contour line passing through their current location. By following this line, they will eventually reach the summit.

However, if there is more than one peak to the mountain, or it has subsidiary hills and ridges, then this approach might yield only a locally optimal result, in that the hikers will simply reach the peak nearest to their starting point. There may be a higher summit elsewhere, but they would be unable to reach it because to do so would entail going down from their current position and descent is forbidden in hill-climbing. One strategy, which is available to hill-climbing computer algorithms, but probably forbidden to even the most athletic hiker, is to leap horizontally from one peak to another in an attempt to move from a slope that would lead only to a local optimum to the one that would lead to the global optimum. Such a procedure is allowed because it does not entail descent. But this might be insufficient because the various slopes may be too widely separated to be reached. Such isolated clusters are referred to as 'islands'. If such islands exist, then one way to maximize the chance of reaching the true highest summit, the global optimum, is to take several randomly chosen starting points and choose that which leads to the highest summit, with or without leaping.

Although the hikers analogy may seem frivolous, its terminology can be translated directly into that of searching for minimum-length cladograms. The highest peak is the set of minimum-length cladograms, the global optimum. The simplest computer algorithms merely make a single pass through the data and construct a single topology. This is equivalent to following the local gradient from where the hikers started. However, although the judicious addition of taxa to the partial cladogram may improve the outcome, the resultant cladogram is most likely to be only locally optimal unless good fortune prevails. More complex routines begin with a single topology, then seek to locate the global optimum by rearranging the clado-gram in various ways. These branch-swapping algorithms are equivalent to jumping between hills. But branch-swapping routines are constrained to try always to decrease the length of the cladogram with which they are currently working. Thus, if the global optimum exists within an alternative set of topologies that can only be reached from the current position by branch-swapping longer cladograms than are currently to hand, then the global optimum can never be reached from that starting point. This is the 'islands of trees' problem. If multiple islands do exist (something that is usually unknown prior to analysis), then we can endeavour to land on the island that includes the most parsimonious solution by running several analyses, each of which starts from a topologically distinct cladogram.

Stepwise addition

Stepwise addition is the process by which taxa are added to the developing cladogram in the initial building phase of an analysis. Initially, a cladogram of three taxa is chosen, then a fourth is added to one of its three branches. A fifth taxon is then selected and added to the network, followed by a sixth and so on, until all taxa have been included. There are various methods for choosing the initial three taxa, the addition sequence of the remaining taxa, and the branch of the incipient cladogram to which each will be added.

The least sophisticated addition sequence selects the first three taxa in the data set to form the initial network and then adds the remaining taxa in the order in which they appear in the data set. The increases in length that would result from attaching a taxon to each branch of the partial cladogram are calculated and the branch selected that would result in the smallest increase. A variation on this procedure uses a pseudorandom number generator to reorder the taxa in the data set prior to cladogram construction. A more elaborate procedure was developed by Farris (1970), which he termed the 'simple algorithm'. First a reference taxon is chosen, usually the first taxon in the data set. Then the difference between this taxon and each of the others is calculated as the sum of the absolute differences between their characters. Farris called this the 'advancement index'. The initial network is then constructed from the reference taxon and the two other taxa that are closest to it, i.e. those that have the lowest advancement indices. The remaining taxa are then added to the developing cladogram in order of increasing advancement index, with ties being broken arbitrarily.

In all of these methods, the order of taxon addition is determined before cladogram construction is begun. In contrast, Swofford (1993) has implemented a dynamic procedure, which he calls 'closest', in which the addition sequence is continually reassessed as the cladogram is built. First, the lengths of the networks for all possible triplets of taxa are calculated and the shortest chosen. Then at each subsequent step, the increase in length that would follow from attaching each of the unselected taxa to each branch of the developing cladogram is calculated and the taxon/branch combination that gives the smallest increase in overall length is chosen. As in all methods, ties are broken arbitrarily. This procedure requires much more computing time than do the other addition sequences. In these, the number of increases in cladogram length that must be calculated at any one step is equal only to the number of possible attachment points (branches). The dynamic procedure multiplies this by the number of unplaced taxa.

No one addition sequence works best for all data sets. The less sophisticated methods are quicker but their inefficiency results in cladograms that may be far from optimal and subsequent branch-swapping may then take longer than it might. Run times using dynamic stepwise addition may be excessive for extremely large numbers of taxa but this is less problematical as processor speeds increase. The random addition sequence is useful in that it can provide a number of different starting points and thereby improve the chances that at least one will lead to the global, rather than a local, optimum.

Random addition can also be employed as a non-rigorous means to evaluate the effectiveness of heuristic procedures. If one runs 100 replicates using random addition and the same set of most parsimonious cladograms are found each time, then one can be reasonably certain that these do represent the set of globally optimal topologies for that data. However, if by the hundredth replication additional topologies or islands are still being discovered, then it is likely that there are even more remaining to be found. Therefore, it is recommended that analyses be repeated several times at least, with the input order of the taxa randomized between runs.

As mentioned above, a major problem with stepwise addition is that one cannot backtrack from a given position. Algorithms with this property are termed 'greedy'. Essentially the problem is their inability to predict the future, that is, which of several options at a given point will ultimately lead to the best result. The placement of a taxon on a partial cladogram may be optimal at that point, but may be seen subsequently to have been suboptimal once further taxa have been added to the network. Once a taxon has been added to a particular branch of a partial cladogram, the consequences of that decision must be accepted. The problem is most acute when ties occur early in stepwise addition. Ties may arise because, at a given stage, the addition of two or more taxa may increase the length by the same minimum amount, or it may be possible to add a single taxon equally to two or more branches, or more than one equally parsimonious topology is found. An 'incorrect' selection may then lead well away from the global optimum. But imagine how our hypothetical hikers could improve their chances of reaching the highest summit if they could climb more than one hill at a time. This is what heuristic programs do when they retain more than one topology at a given step. These may simply be the set of shortest partial cladograms at that stage, but a fixed number may also be selected. Then, suboptimal topologies may also be retained. This procedure reduces the effect of ties because, to some extent (depending upon the number of topologies retained), each of the alternatives is followed up.

Branch-swapping

In practice though, manipulation of addition sequence alone will generally yield only a local optimum. However, it may be possible to improve on this by performing a series of predefined rearrangements of the cladogram, in the hope that a shorter topology will be found. These rearrangements, commonly referred to as 'branch-swapping', are very hit-and-miss, but if a shorter topology does exist and sufficient rearrangements are performed, then one of these rearrangements is likely to find it.

Branch-swapping algorithms are implemented by all cladistic computer packages. The simplest rearrangement is nearest-neighbour interchange (NNI), sometimes referred to as local branch-swapping. Each internal branch of a bifurcating cladogram subtends four 'nearest-neighbour' branches, two at either end. In Fig. 3.3, these are A + B, C, D, and E + F. NNI then exchanges a branch from one end of the internal branch with one from the other, e.g. C with E + F. For any internal branch, there are just two such NNI rearrangements. The procedure is then repeated for all possible internal branches and the lengths of the resulting topologies calculated to determine whether they are shorter.

More extensive rearrangements can be performed and such methods are often referred to as global branch-swapping. These involve clipping the cladogram into two or more subcladograms and then reconnecting these in various ways, with all possible recombinations being evaluated. 'Subtree pruning and regrafting' (SPR) clips off a rooted subcladogram (Fig. 3.4) from the main cladogram. This is then regrafted on to each branch of the remnant cladogram in turn and the length of the resultant topologies calculated. All possible combinations of pruning and regrafting are evaluated. In contrast, in 'tree bisection and reconnection' (TBR) (Fig. 3.5), the clipped subcladogram

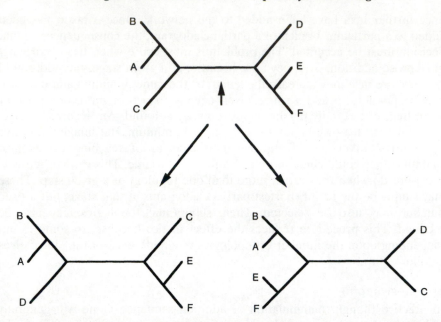

Fig. 3.3 Example of branch-swapping by nearest-neighbour interchange. The chosen branch (indicated by the short arrow) has a pair nearest-neighbour groups at either end (A + B, C; D, E + F). One of these from the left end (C) is interchanged with each of those on the right end (D or E + F) in turn to give two new topologies. (After Swofford and Olsen 1990).

is re-rooted before it is reconnected to each branch of the remnant clado-gram. All possible bisections, re-rootings and reconnections are evaluated. The SPR and TBR routines of PAUP cut the cladogram into only two pieces, while PIWE and NONA permit the cladogram to be cut into a maximum of ten pieces.

The effectiveness of these branch-swapping routines in recovering the optimum set of cladograms increases in the order: NNI, SPR, TBR. However, the more rigorous the branch-swapping method applied, the more compar-isons there are to evaluate and the longer is the computational time required. For large data sets, it may not be possible for analyses using TBR to be completed within an acceptable time. As with many aspects of cladistic analysis, a trade-off must be made between confidence in having obtained the global optimum and computation time. Several algorithms to speed up branch swapping searches for most parsimonious cladograms were described by Goloboff (1996).

However, like stepwise addition, branch-swapping suffers from problems of becoming trapped in local optima. Unless there is an unbroken series of rearrangements between the initial cladogram and the minimum-length topology, even branch-swapping routines will not find the global optimum.

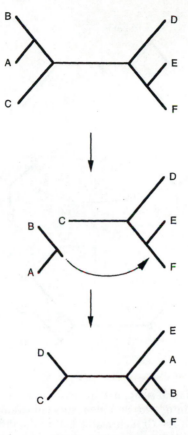

Fig. 3.4 Example of branch-swapping by subtree pruning and regrafting. A rooted subcladogram, A + B, is clipped from the main cladogram then reattached to another branch (leading to taxon F) to give a new topology. (After Swofford and Olsen 1990).

For example, reaching the most parsimonious cladogram may require passing through a series of rearrangements each of which is the same length as the preceding. If the current cladogram is replaced only if the new topology is shorter, rather than simply being of the same length, then crossing these 'plateaux of optimality' will not be possible. The solution to this problem is to retain all the most parsimonious solutions found during a given round of rearrangements. Equally, if reaching the global optimum requires rearrangement of cladograms that are longer than the best found so far, entrapment in a local optimum will also occur. Again, the solution is to retain more than one cladogram, but now to include suboptimal topologies as well, in the expectation that one will lead to the global optimum. But even TBR will fail to lead from an initial to the optimal cladogram if the differences between the two, which may be quite minor, are in separate parts of the cladograms

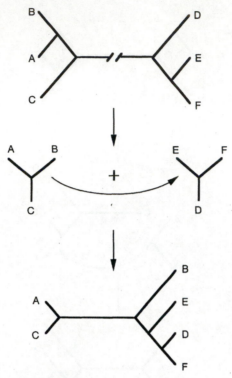

Fig. 3.5 Example of branch-swapping by tree bisection and reconnection. The cladogram is divided into two unrooted subcladograms. One subcladogram is rerooted (between B and A + C) then reattached to the other subcladogram (on the branch leading to taxon E) to give a new topology. (After Swofford and Olsen 1990).

(Page 1993*a*). Multiple cutting, as implemented in PIWE and NONA, is an improvement in this regard, but again, for large data sets, application of such comprehensive options may stretch computation time beyond acceptable limits.

3.2 CHARACTER POLARITY AND ROOTING

Polarization refers to the imposition of direction onto character state change or character transformation. A character is said to be polarized when the plesiomorphic state has been distinguished from the apomorphic state. Manually implemented cladistic methods, such as Hennigian argumentation, require apomorphies to be determined in advance of cladogram construction and thus characters must be polarized *a priori*. Numerous methods and criteria for identifying the apomorphic states of characters have been proposed and classified in various ways. However, the classification advanced by

Nelson (1973) identifies what is perhaps the most fundamental division. Nelson recognized as 'indirect' those methods that require information from a source external to the study taxa. Most often, this is the prior existence of a higher-level phylogeny including the study taxa. Because the principle of parsimony is fundamental to the application of outgroup comparison, Nelson considered this to be the only valid indirect method. However, the higher-level phylogeny itself must be justified in terms of one that is even more inclusive, and so on, leading to an infinite regress. Eventually, recourse must be made to a method that is independent of pre-existing hypotheses of relationship in order to validate outgroup comparisons. Nelson termed such methods direct arguments because they can be implemented using only the information available from members of the study group. Only the criterion of ontogenetic character precedence was considered by Nelson to be valid because it did not rely on *a priori* models of character transformation.

3.2.1 Outgroup comparison

In its simplest form, the outgroup criterion for polarity determination is defined as follows.

For a given character with two or more states within a group, the state occurring in related groups is assumed to be the plesiomorphic state (Watrous and Wheeler 1981: 5).

Prior to 1980, application of outgroup comparison had been inconsistent and confusion with ingroup commonality (see below) was widespread. Watrous and Wheeler provided initial clarification with the formulation of a set of operational rules for outgroup comparison that they termed the 'functional ingroup/functional outgroup' (FIG/FOG) method. First, an outgroup to the unresolved set of ingroup taxa is designated. Then, a character is selected and through comparison with the state occurring in the outgroup, partial resolution of the ingroup is achieved. This initial character is chosen so as to resolve the basal node of the ingroup. The first taxon to branch from the ingroup topology is then treated as the functional outgroup to the remaining ingroup taxa (the functional ingroup), thereby permitting further resolution of the ingroup. This procedure is repeated until full resolution of the ingroup has been achieved. An example of the FIG/FOG method is shown in Fig. 3.6.

The FIG/FOG approach to outgroup comparison is adequate when the states displayed in the outgroup are invariant. However, it cannot be applied easily or consistently if there is variation in the outgroup, and particularly if one or more outgroup taxa possess a state that is present also within the ingroup. The problem of a heterogeneous outgroup was first addressed in detail by Maddison *et al.* (1984). They noted that determining the plesiomorphic state of a character solely by reference to the members of the ingroup

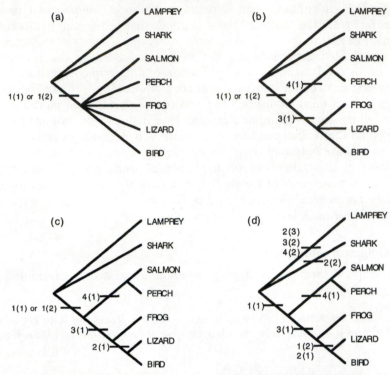

Fig. 3.6 Example of the FIG/FOG method of outgroup comparison, using the data in Table 3.3, in which an ingroup of six gnathostomes and one outgroup, the lamprey, are scored for four unordered, multistate characters. (a) The shark is established as the first functional outgroup because it shares state 0 of character 1 with the lamprey. Consequently, this state is interpreted as plesiomorphic for character 1. However, while the remaining five ingroup taxa have been shown to form a monophyletic group, it remains uncertain whether this clade, the Osteichthyes, is supported by state 1 or state 2 of character 1. (b) Next, using the shark as functional outgroup to the Osteichthyes and ignoring the lamprey, state 1 of character 3 is found to be apomorphic for tetrapods (frog, lizard and bird) and state 1 of character 4 to be apomorphic for bony fish (salmon and perch). (c) Finally, using the bony fish as functional outgroup to the tetrapods, state 1 of character 2 is seen to unite the two amniotes (lizard and bird) into a monophyletic group that excludes the frog. (d) The relative apomorphy of states 1 and 2 of character 1 can now be resolved, with state 1 supporting the monophyly of the Osteichthyes and state 2 the monophyly of the Amniota, and the outstanding autapomorphic states placed onto the fully resolved cladogram.

would frequently lead only to locally parsimonious solutions. Failure to achieve global optimality might also result if outgroup comparison was taken to indicate the character state present in the most recent common ancestor of the ingroup. These two defective procedures are often combined to produce the 'groundplan estimation' method, which despite being methodologically

Table 3.3 Data set to illustrate the FIG/FOG method
of outgroup comparison

Taxon	Character 1 2 3 4
Lamprey	0 3 2 2
Shark	0 2 0 0
Salmon	1 0 0 1
Perch	1 0 0 1
Frog	1 0 1 0
Lizard	2 1 1 0
Bird	2 1 1 0

flawed (Nixon and Carpenter 1996*a*), is still strongly advocated in certain
quarters. Similarly, if the state found most commonly among the outgroup
taxa is taken to be plesiomorphic, then again only a local optimum may be
found, depending upon the interrelationships of the outgroup taxa and the
distribution of the various character states among them.

Rather than estimating the condition present in the most recent common
ancestor of the ingroup (the 'ingroup node'; Fig. 3.7), Maddison *et al.* argued
that it is the state at the next most distal node (the 'outgroup node'; Fig. 3.7)
that must be determined if the global optimum is to be found. This is the
node that unites the ingroup and the first outgroup into a monophyletic unit.
The state assigned to the outgroup node is termed 'decisive' if it can assume
only a single most parsimonious value, or 'equivocal' (or ambiguous) if more
than one value may be applied. Furthermore, relationships among the outgroup taxa are assumed to be both known and fixed.

Visual inspection is sufficient to resolve the simplest cases. For example, if
the outgroup taxa all agree in the state they possess, then this state is the
decisive assignment at the outgroup node. Conversely, if the first two outgroup taxa differ in their states, then the assignment at the outgroup node is
equivocal. Additional, more basal outgroup taxa can never convert the
assignment from equivocal to decisive, demonstrating its independence from
the frequency of the states among the outgroup taxa.

However, visual inspection can fail to recover the most parsimonious

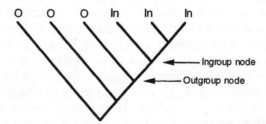

Fig. 3.7 Illustration of the ingroup and outgroup nodes. Outgroup taxa are labelled
O; ingroup taxa are labelled In.

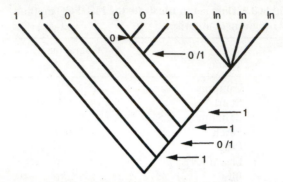

Fig. 3.8 Example of the algorithmic approach to outgroup comparison, applied to seven outgroup taxa and an unresolved set of four ingroup taxa (In). The procedure is similar to Wagner optimization (see §4.1.1). First, the outgroup taxa are labelled with their states, 0 or 1 (if an outgroup is polymorphic, it is labelled 0/1). Then, beginning with pairs of outgroup taxa and proceeding towards the outgroup node (ignoring the root), the internal nodes are labelled according to the following rules: a node is assigned state 0 if its two derivative nodes are both labelled 0, or are 0 and 0/1; a node is assigned state 1 if its two derivative nodes are both labelled 1, or are 1 and 0/1; a node is assigned state 0/1 if its two derivative nodes are both labelled 0/1, or are 0 and 1. Here, the outgroup node is decisively assigned state 1.

reconstruction for more complex examples and an algorithmic approach is required (Fig. 3.8). The properties of this algorithm lead to further rules that permit the outgroup node state to be determined by simple inspection of the distribution of states in the outgroup. When two successive outgroup taxa share the same state (i.e. they form a 'doublet'), if this is also the state found in the first outgroup, then it is decisive for the outgroup node (Fig. 3.9a). Should the first doublet and the first outgroup taxon disagree, then the state at the outgroup node is equivocal (Fig. 3.9b). Note that all outgroup structure beyond the first doublet is irrelevant to the assessment. If there are no doublets (i.e. the states in each successively more distal outgroup alternate), then if the most distal outgroup agrees in state with the first outgroup, this state is decisive at the outgroup node (Fig. 3.10a). Otherwise, the assessment is equivocal (Fig. 3.10b). Clearly, where there is no doublet, the choice of the most distal outgroup is critical and the addition or subtraction of just one outgroup to the base of the cladogram would be sufficient to change an outgroup node assessment from decisive to equivocal or vice versa. However, such a scenario is most unlikely in practice. If a pattern of alternating character states were to be found in a large outgroup, the character would most probably be rejected as having very low information content.

In the systematic literature, great, even paramount, importance is some-times attached to the first outgroup taxon, that is, the sister group of the ingroup. If the sister group cannot be identified with certainty, it is thought that characters cannot be polarized and resolution of the ingroup is impossi-

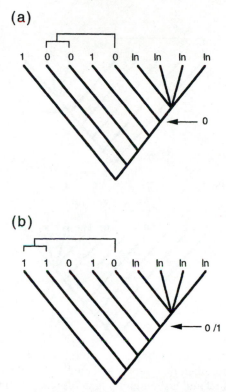

Fig. 3.9 Illustration of the 'first doublet rule' for a binary character. (a) If the state in the first doublet agrees with that in the first outgroup taxon, then this state is assigned decisively to the outgroup node. (b) If the state in the first doublet disagrees with that in the first outgroup taxon, then the state assigned to the outgroup node is equivocal.

ble. In fact, identifying the sister group is not so important. It is true that this taxon plays a major role, because its state will always be assigned to the outgroup node, either decisively or equivocally, but more distal outgroup taxa also exert an effect. In contrast, use of the sister group to polarize characters is sometimes criticized if this taxon is considered to be too 'derived', that is, it has too many autapomorphies, to make comparisons with the ingroup meaningful. However, this is no justification for ignoring it and appealing to some more distant and supposedly more 'primitive' taxon.

More important are the preconditions of both the FIG/FOG and algorithmic methods that the relationships among the outgroup taxa are both specified and fixed. Problems may arise when the interrelationships among the outgroup taxa are partially or even wholly unresolved. Then, uncertainty in outgroup relationships is translated into uncertainty in the state assignment at the outgroup node. However, this is not the problem that it may first

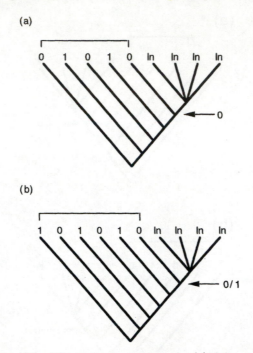

Fig. 3.10 Illustration of the 'alternating outgroup rule'. (a) If the states of the first and last outgroup taxa agree, then this state is assigned decisively to the outgroup node. (b) If the state in the first and last outgroup taxa disagree, then the state assigned to the outgroup node is equivocal.

appear, for we will argue later in this chapter (§3.2.5) that character polarity is actually a property that is derived from a cladistic analysis, rather than an *a priori* condition.

3.2.2 The ontogenetic criterion

In the analysis of ontogeny, Løvtrup (1978) recognized two types of character, based upon their developmental interactions. Epigenetic characters are causally related among themselves, such that each stage in ontogeny is a modification of, or is induced by, another character that developed earlier in the ontogenetic sequence. Non-epigenetic characters are not so causally related. Epigenetic characters are generally viewed as fundamental to and essential for normal morphogenesis. The notochord of vertebrates is an epigenetic character because normal vertebrate development cannot proceed in its absence. In contrast, the presence of pigments in the feathers of a bird is non-epigenetic because it is quite possible to have a viable bird, an albino, that lacks all such pigmentation. Løvtrup further distinguished between terminal characters, which occur last in an ontogenetic sequence, and non-terminal characters, which are those that occur earlier in the sequence.

Original sequence:	$X \rightarrow Y \rightarrow Z$	
Addition	$X \rightarrow Y \rightarrow Z \rightarrow D$	$X \rightarrow Y \rightarrow E \rightarrow Z$
Substitution	$X \rightarrow Y \rightarrow F$	$X \rightarrow G \rightarrow Z$
Deletion	$X \rightarrow Y$	$X \rightarrow Z$

Fig. 3.11 The six fundamental ways by which an ontogenetic pathway can be modified.

Furthermore, an ontogenetic sequence can undergo three fundamental types of modification: addition, deletion and substitution, each of which can be applied to both character types, giving a total of six pathways by which an ontogeny can change (Fig. 3.11).

Historically, characters drawn from ontogeny have been applied to phylogenetic reconstruction in two ways. The simplest viewpoint is that of Haeckel, developed in the 1860s, whose Biogenetic Law states simply that ontogeny recapitulates phylogeny. Under this interpretation, the ontogenetic development of a species passes through stages displayed in the adults of its ancestors. Ontogeny is allowed only to proceed by a highly restricted form of terminal addition, in which characters are added to the end of an already completed ontogenetic sequence. Thus the course of phylogeny can be discovered simply by 'reading' the ontogenetic sequence. However, this interpretation of recapitulation has long been rejected as too strict and simplistic. Nevertheless, the idea that phylogeny could somehow be 'read' from ontogenetic sequences remains.

An alternative viewpoint had already been proposed in the 1820s by the German comparative embryologist, von Baer, who summarized the results of his studies as follows.

- In development, the general characters appear before the special characters.

- From the more general characters, the less general and, finally, the special characters are developed.

- During its development, an organism departs more and more from the form of other organisms.

- The early stages in the development of an organism are not like the adult stages of other organisms lower down on the scale, but are like the early stages of those organisms.

Although von Baer (1828) did not frame his rules in an evolutionary context, he did make two great contributions towards forging a strong link between orderly ontogeny and phylogenetic inference. The first was his recognition that we would never expect the ontogenetic sequence of an organism to pass through the stages found in the adults of its ancestors.

Rather, during ontogeny, two taxa would follow the same course of development up to the point at which they diverged into separate lineages. Both ontogenies would then be observed, in general, to have undergone one or more independent terminal substitutions or additions, depending on the time that had elapsed since differentiation and the amount of subsequent change that had taken place.

But perhaps more important was his second rule, which stated that ontogenetic change proceeds from the more to the less general. This observation was generalized by Nelson (1978: 327) into the following definition of the ontogenetic criterion for determining character polarity.

Given an ontogenetic character transformation, from a character observed to be more general to a character observed to be less general, the more general character is primitive and the less general character advanced.

The application of the ontogenetic criterion can be illustrated using the example of the vertebrate endoskeleton. The endoskeleton of the adult shark is composed of cartilage, while that of the perch is largely made out of bone. Given only these observations, no decision can be made as to whether cartilage or bone is apomorphic. However, a study of the ontogeny of the two taxa shows that while a cartilaginous endoskeleton is formed in the early embryos of both taxa, only in the shark does this state persist into the adult animal. In contrast, later in the ontogeny of the perch, the cartilage is largely replaced by bone, which is the state found in the adult. In other words, a character that is observed to be more general (cartilage) has transformed into one that is observed to be less general (bone), from which it can be inferred that a cartilaginous endoskeleton is plesiomorphic and a bony endoskeleton apomorphic.

Alberch (1985) and Kluge (1985) disagreed with this interpretation of the ontogenetic process, arguing that the valid ontogenetic characters were not the observed states but the transformational processes between those states. Thus, in the above example, there would be only a single ontogenetic character, the transformation of a cartilaginous skeleton into a bony one. De Queiroz (1985) objected to observed states in an ontogenetic sequence being used as characters in cladogram construction, because these 'instantaneous morphologies' were abstractions from 'real' ontogenetic transformations. In his view, phylogeny is a sequence of never-ending life cycles and thus the evidential basis for inferring phylogeny should be the ontogenetic transformations themselves, rather than the features of the organisms that were being transformed. As a result, he concluded that there could be no 'ontogenetic method'.

Kluge (1988) considered de Queiroz's concept of treating transformations as characters both to be incomplete and to offer no advantage over describing the life cycle in terms of a model of growth and differentiation (Kluge and

Strauss 1985). Furthermore, taken to the extreme, de Queiroz's approach would reduce the entire organism, if not the entire living world, to a single character with an immense number of states (transformations) and there would be no basis for comparative biology. In practice, de Queiroz adopted a pragmatic approach, defining 'character' as a feature of an organism 'large enough to encompass variation that is potentially informative about the relationships among the organisms being studied'. However, this definition is open to the criticism of how large is large and how one is expected to determine what is 'potentially informative' prior to conducting an analysis.

Further controversy regarding Nelson's generalization of the ontogenetic criterion concerned what exactly is meant by 'general', for which there are two interpretations:

- strict temporal precedence, so that the more general state is that which occurs first in ontogeny; and

- the most frequently observed, so that the general state is the commonest state.

If the latter definition is adopted, then Nelson's criterion is nothing more than a special case of ingroup commonality, with all the deficiencies associated with that method (see below). However, most cladists would employ the first definition, under which general is something more than simply most commonly observed. Although the more general character will be more common, insofar as it will be possessed by all those taxa that also show the less general character as well as some others that do not, this does not equate to ingroup commonality (see §3.2.4). Frequent occurrence bears no necessary relationship to relative time of phylogenetic appearance. A character may indeed be more common than another but what is important is that the less general character is nested within the distribution of the more general character (Weston 1994). Without such an unequivocal relationship of generality, the more common character will not be the more general.

Furthermore, equating generality to strict temporal precedence results in Haeckel's Biogenetic law. However, as already noted, this interpretation of ontogeny is too strict because it only operates if ontogenetic change occurs by terminal addition. This led Løvtrup and others to assert that Nelson's definition would only apply to ontogenies that were modified in this way. Sequences in which characters have been added or deleted subterminally could not be analysed under Nelson's definition. Should characters be substituted terminally, then information would be lost and outgroup comparison would have to be used to resolve polarity.

Terminal deletion has been considered to pose a particular problem for Nelson's criterion because it gives rise to secondarily simplified ontogenies that cannot be distinguished from plesiomorphic, unmodified ontogenies.

Paedomorphosis, or phylogenetic neoteny, in which a feature that appears only in the juveniles of ancestors occurs as both a juvenile and an adult character in descendent taxa, results from such terminal deletion. The occurrence of paedomorphosis falsifies von Baer's law because the ontogeny cannot be interpreted as going from the more general to the less general and thus Nelson's criterion is inapplicable. Of course, it is possible for any sequence in which the more general character is retained throughout ontogeny to be due to paedomorphosis rather than the retention of the plesiomorphic condition. But this is no more than an ontogenetic restatement of a problem that pervades all systematics, that is, the detection of homoplasy. As noted in Chapter 1, errors due to the interpretation of homoplasy as synapomorphy are detected by character incongruence and resolved by the application of parsimony. In this regard, paedomorphosis is treated no differently from any other instance of homoplasy.

For example, when we examine the ontogeny of certain salamanders, we observe that the juvenile cartilaginous skeleton remains largely unaltered through to the adult stage. Using Nelson's criterion alone, we cannot differentiate this ontogeny from the plesiomorphic sequence observed in the shark. However, analysis of other characters shows that salamanders are deeply embedded within the Osteichthyes, one generally acknowledged apomorphy of which is a bony endoskeleton in the adult. Thus, congruence with other characters shows that the cartilaginous skeleton of these salamanders is due to paedomorphosis.

However, Weston (1988) showed that it was possible to hypothesize situations that did not conform to the requirement for terminal addition, but which nonetheless could still be analysed using a more broadly framed direct method of character analysis. Under Nelson's Law, the direction of ontogenetic transformation need not necessarily be viewed as an indicator of the direction of phylogenetic transformation (Weston 1994). Rather, it is a criterion of similarity, like positional correspondence, and is thus a necessary but not sufficient criterion for establishing primary homology. Weston sought to remove any reference to sequence in the direct method by replacing the concept of ontogenetic transformation with the more general concept of homology and emphasizing the role of directly observed generality relationships between homologous characters. He was thus able to further generalize Nelson's Law (Weston 1994: 133):

Given a distribution of two homologous characters in which one, x, is possessed by all of the species that also possess its homolog, character y, and by at least one other species that does not, then y may be postulated to be apomorphous relative to x.

From this standpoint, the only valid information that can be extracted from ontogenetic transformations is the relative generality of characters. However, de Pinna (1994) regarded the information contained in the order in which characters transform one into another to constitute ontogenetic information *per se*. To ignore this orderliness was to overlook the essential systematic information derivable from ontogenies. He considered that eliminating the relevance of ontogenetic sequence information reduced the direct method to a form of 'common equals primitive' procedure. However, as noted above, this would only hold if the more general character is equated strictly with the commonest character, a correspondence specifically rejected by Weston, who emphasized the hierarchical nesting of characters. It would thus seem that de Pinna misunderstood the concept of generality as it is applied in both Nelson's Law and Weston's generalization.

3.2.3 Ontogenetic criterion or outgroup comparison—which is superior?

Once the ontogenetic criterion and outgroup comparison had both been formally defined, and earlier errors in interpretation and implementation corrected, there was a period of debate over their relative merits. One defect of outgroup comparison was perceived to be the requirement that the relationships among the outgroup taxa be prespecified. In the absence of such an hypothesis, outgroup comparison was considered difficult at best and at worst, totally impractical. However, because the hierarchical relationship between plesiomorphic and apomorphic characters could be observed directly, the ontogenetic criterion was considered independent of higher level hypotheses of relationship and therefore superior. Proponents of outgroup comparison responded by observing that the ontogenetic criterion could only be applied in cases where the ontogenetic sequence changed by terminal addition, and thus would be misled by paedomorphosis. In order to polarize characters in which paedomorphosis had occurred, recourse had to be made to outgroup comparison. This argument was answered with the observation that paedomorphosis was a problem for systematics in general rather than ontogeny in particular. Nevertheless, it was counter-claimed, because the direct observation of ontogeny could not polarize any character that could not also be polarized by outgroup comparison, and because there were instances where the ontogenetic criterion failed but outgroup comparison succeeded, then the former was merely an incomplete version of the latter.

However, much of this debate is essentially beside the point because it viewed the ontogenetic criterion and outgroup comparison either as competing alternatives or as essentially the same. Both perspectives are erroneous. Despite their difference of opinion as to what constitutes information in ontogenetic data, both Weston (1994) and de Pinna (1994) agreed that the ontogenetic criterion and outgroup comparison are complementary with non-overlapping roles. Thus, empirical studies to determine the comparative worth of the two methods are futile. In itself outgroup comparison does not polarize characters, nor is it a means for rooting cladograms (Weston 1994). It is actually a method for simultaneously constructing the most parsimonious cladogram for the ingroup and locating that cladogram within the larger scheme encompassing all living organisms. Outgroup comparison should be treated as a technique for rooting these partial cladograms of study taxa, and hence polarizing characters, on the generally reasonable assumption that the root of all life lies outside the ingroup in question.

Thus, in a sense, outgroup comparison can be characterized as providing only 'local rooting'. Global rooting of the entire tree of life would require the application of a direct method, such as the ontogenetic criterion. However, only if we were working at the very base of the tree of life and were fortunate enough to discover the appropriate ontogenetic transformation, would we be able to polarize the entire topology. Until that time, even the ontogenetic criterion provides only local polarity determination. Thus, for all practical purposes, the ontogenetic criterion is little different from outgroup comparison. Both approaches are valid because both are ultimately justified by parsimony.

3.2.4 *A priori* models of character state change
Ingroup commonality
In addition to the ontogenetic criterion and outgroup comparison, there are many other criteria for determining character polarity that are based upon models of how character states are believed to transform one to another. Transformational models are frequently unjustified by any theoretical or empirical framework. For example, the criterion of ingroup commonality is defined as follows:

The plesiomorphic state will be more widespread within a monophyletic group than will any one apomorphic state. Therefore, the state occurring most commonly within the ingroup is plesiomorphic.

Put simply, common equals primitive. In this form, ingroup commonality is *ad hoc* because it assumes that the evolutionary process is conservative and that plesiomorphic characters are likely to be retained. While this may be true for a particular character and group of taxa, it is not necessarily always true. To distinguish between these two cases, outgroup comparison is needed and ingroup commonality becomes superfluous. Consider character 1 in Table 3.3, in which the lamprey is regarded as the outgroup. Using ingroup commonality, state 1 would be interpreted as plesiomorphic, with state 0 being autapomorphic for the shark and state 2 uniting the lizard and bird. State 0 in the lamprey would be ignored as this taxon does not constitute part of the ingroup. However, using outgroup comparison, state 0 is interpreted as plesiomorphic and state 1 apomorphic, with subsequent transformation into state 2.

There is an even more fundamental flaw in ingroup commonality. The basic component of cladistic analysis is the three-taxon statement, that is, taxa A and B are more closely related to each another than either is to a third taxon, C. Resolution of a three-taxon statement depends upon the grouping information (apomorphy) being present in A and B and absent in C. However, using ingroup commonality, a three-taxon statement can never be recognized, let alone resolved, because this requires two of the three taxa share the apomorphic state. Ingroup commonality is thus contrary to the very basis of cladistic analysis.

Stratigraphy

Fossil taxa are often held to be of paramount importance in determining character polarity using the stratigraphic criterion or the criterion of geological character precedence, which states:

If one character state occurs only in older fossils and another state only in younger fossils, then the former is the plesiomorphic and the latter the apomorphic state of that character.

The validity of the general concept is not in doubt; plesiomorphy must precede apomorphy in time. However, equating the condition observed in the oldest fossil with the plesiomorphic condition is fraught with difficulties. We are required to assume that the available sequence of fossils is a true and accurate record. However, this is rarely demonstrably so. Furthermore, there is no reason to assume that just because one fossil is older than another that *all* its characters are therefore plesiomorphic. There are many examples of very old fossil taxa with many apomorphic features. Many soft-bodied groups of organisms had very early origins but left no fossil evidence. In contrast, many hard-bodied organisms that arose later left excellent fossil records from the time they appeared. Strict application of the stratigraphic criterion would lead to incorrect conclusions in such circumstances.

At present the stratigraphic criterion has few strict adherents. Using it to establish character polarity is always suspect, for any inconsistent or unwanted results can be explained away by invoking, *ad hoc*, the incompleteness of the record. Stratigraphy may be useful, however, in allowing us to choose among multiple, equally most parsimonious cladograms, on the basis of concordance with the fossil record.

Biogeography

Several criteria have been proposed that use biogeographical information to polarize characters. The most widely known is the 'criterion of chorological progression' (or 'progression rule', Hennig 1966), which postulates that the most derived species will be that found furthest geographically or ecologically from the ancestral species. However, as a method for inferring character polarity, the progression rule suffers several deficiencies. In particular, even if allopatric speciation occurs, it cannot be assumed that the character states found only in the peripheral populations are apomorphic. Vicariance bio-geography is neutral regarding character polarity. Furthermore, evidence is required of the historical distribution or ecological requirements of the ancestral species. However, given that ancestral taxa cannot be unequivocally recognized, such evidence will not be forthcoming. Such arguments invalidate the progression rule as a means of inferring character polarity. Most cladists would choose to use cladograms to test biogeographical hypotheses. If such tests are to be independent, then cladograms must be constructed without recourse to biogeographical information.

Function/adaptive value

The hypothesized functional or adaptive value of a character is also fre-quently held to be of fundamental importance in polarity determination. Functional arguments are usually couched in terms of niche restriction or specializations that help an organism out-compete its relatives. But many authors conflate functional value with selective value. Such usage is fraught with difficulties, not the least of which is how the nature of the selective forces acting on a character are to be measured. In fact, most studies that purport to use function to polarize characters do nothing of the sort (Lauder 1990). They simply present morphological data, then infer function, not measure it. The correct interpretation of functional data requires that the use of structural characters by the organism be directly observed and quantified. But such functional characters have no special properties that single them out as intrinsically superior. Functional characters can be admitted into a cladistic analysis but functional considerations should not be used *a priori* to determine polarity.

Underlying synapomorphy

Perhaps the most idiosyncratic model of character change is that of underly-ing synapomorphy. Championed by Ole Saether, underlying synapomorphy is defined as 'close parallelism as a result of common inherited genetic factors causing incomplete synapomorphy' or 'the inherited potential to develop parallel similarities', although this potential may not be realized in all descendants. Consequently, it refers to the occurrence of synapomorphies in only some members of a putative monophyletic group. Consider a highly simplified cladogram of the Bilateria (Fig. 3.12). Haemoglobin is known to occur in only three of the many lineages: *Tubifex* worms, some chironomid midges and vertebrates. The application of standard optimization methods (see below) would lead us to conclude that haemoglobin had been indepen-dently derived three times. However, the amino-acid sequences in the three groups are all very similar and it would seem most unlikely that such a complex molecule could have arisen *de novo* on three separate occasions. Underlying synapomorphy would assert that we are mistaken when we code the other taxa as lacking haemoglobin. These taxa all possess the relevant

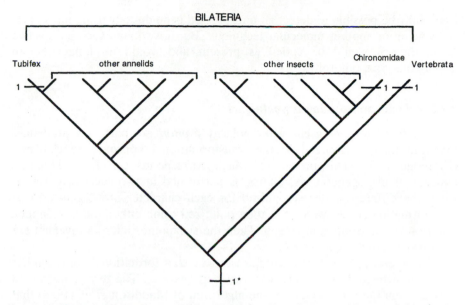

Fig. 3.12 Underlying synapomorphy asserts that the presence of haemoglobin (indicated by 1) in *Tubifex* annelids, certain chironomid midges and vertebrates implies support for the monophyly of the entire Bilateria (1*). The lack of observable haemoglobin in all other Bilateria is not taken as evidence against this hypothesis for it is argued that these taxa possess the unexpressed capacity to develop this molecule.

genes, but the genes are switched off because it is simply not selectively advantageous for them to be active. It is this unexpressed capacity to develop a feature that is the underlying synapomorphy. Thus, in this example, the potential 'capacity' (expressed or not) to manufacture haemoglobin can be used to unite all the taxa in Fig. 3.12 into a monophyletic group.

Saether argued that the use of underlying synapomorphies had been advocated by Hennig (1966), who called them homoiologies. However, Hennig considered homoiology to be equivalent to convergence, which he specifically rejected as a valid tool for estimating cladistic relationships. It may indeed be true that the haemoglobin gene occurs in all taxa in Fig. 3.12. However, the monophyly of the Bilateria cannot be supported by only scattered occurrences of an expressed gene, for this is using the absence of information (observations, characters) as evidence for groups. By invoking *ad hoc* hypotheses to explain away conflict in the data as 'unexpressed', underlying synapomorphies provide a licence to group in any way whatsoever. While it is perfectly acceptable to use the occurrence of haemoglobin in *Tubifex* worms, chironomid midges and vertebrates as evidence of the monophyly of each group separately, the distribution of this character does not provide evidential support for the monophyly of the Bilateria as a whole. It may eventually be possible to detect an unexpressed gene directly using nucleotide sequencing or another molecular technique. But if we found such an inactive gene, then it would be coded as present and would no longer be an underlying synapomorphy.

3.2.5 Polarity and rooting *a posteriori*

Criteria for determining character polarity *a priori* are part of a procedure that can be characterized as the transformational approach to cladistics (Eldredge 1979). This procedure, which corresponds closely to Hennig's concept of phylogenetic systematics, is performed in two successive stages. First, the different states recognized for each character are organized into transformation series, which are then polarized using either the ontogenetic criterion or outgroup comparison. Then the synapomorphies so revealed are used to construct a cladogram.

With reference to outgroup comparison, the transformational approach has been characterized as 'constrained, two-step analysis'. The term constrained derives from the requirement of the algorithm of Maddison *et al.* (1984) that the outgroup must be resolved *a priori* and those relationships then held fixed during polarity determination and cladogram construction. However, constrained analysis requires that two *a priori* assumptions be made. The first is that the ingroup taxa form a monophyletic group, which in turn, implies that the root must be basal to the ingroup and not within it. The second assumption concerns the outgroup structure, which implies fixed hypotheses of monophyly, both among the outgroup taxa and with respect to the ingroup. However, these immutable patterns of relationship are not open to independent testing and, in particular, none of the outgroup taxa is permitted to be part of the ingroup.

Problems arise when there is parallel homoplasy between one or more

---

▷

Fig. 3.13 Outgroup constrained and simultaneous, unconstrained analysis of three ingroup taxa (A–C) and three outgroup taxa (D–F). (a) With outgroup relationships predetermined and constrained as shown, character 5 is interpreted as synapomorphic for group A–E, but with secondary loss in taxon A + B. The total length of the cladogram is six steps. (b) Two more characters, 6 and 7, are added that have the same distribution among taxa as character 5. Because the outgroup relationships are fixed, then these are also forced to show secondary loss in taxon A + B. The cladogram now has ten steps. (c) However, if the constraint on outgroup relationships is removed and all the taxa analysed simultaneously, then three shorter cladograms of nine steps are found in which the original ingroup is not monophyletic. It can also be noted that the cladogram in which taxa C, D and E form a trichotomy is both the strictly supported cladogram (see §4.2) and also the strict consensus tree of the three cladograms (see §7.2).

members of the outgroup and a subset of the ingroup. For example, consider a cladogram (Fig. 3.13a) with three ingroup taxa (A, B and C) and three outgroup taxa (D, E and F), the relationships among which are determined by five characters, 1–5. Four of these characters are unique and unreversed apomorphies but character 5, while offering some support to clade A–E, is secondarily lost in taxon A + B. The cladogram is thus six steps long. Subsequently, if two new characters, 6 and 7, are found that have the same

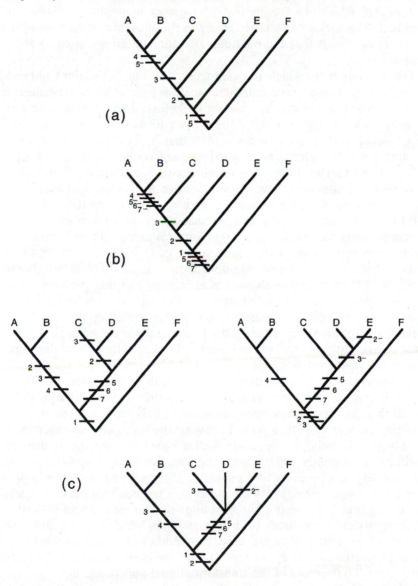

distribution among the taxa as character 5, then because the outgroup relationships are fixed, outgroup comparison must yield the same topology as before (Fig. 3.13b), now with ten steps. However, if all seven characters are analysed without topological constraints, then three equally parsimonious cladograms result (Fig. 3.13c) (see §4.2 for the reasons why there should only be a single solution). These cladograms are each one step shorter than that in Fig. 3.13b, but more importantly they also have a markedly different topology and one in which the assumption of ingroup monophyly is found to be incorrect. This lack of global parsimony in the presence of homoplasy is due to the requirement that the outgroup structure be known *a priori* and then held invariant.

The procedure by which the cladograms in Fig. 3.13c were obtained is termed 'simultaneous, unconstrained analysis'; simultaneous because both outgroup and ingroup taxa are analysed together, unconstrained because the outgroup relationships are unspecified prior to analysis. This approach can never yield less parsimonious cladograms than a two-step, constrained analysis and will often give a more parsimonious result. Nixon and Carpenter (1993) stated that in order for a simultaneous, unconstrained analysis to be fully effective, characters that are informative with respect to outgroup taxa relationships must also be included, even if these characters are invariant within the ingroup. However, this is unnecessary. In a given analysis, we are concerned only with resolving the relationships among the ingroup taxa. The interrelationships of the outgroup taxa are quite irrelevant. If we are interested in this more inclusive question, then a separate data set should be formulated that comprises characters appropriate to that problem. If we do exclude from the data set all characters relevant to resolving the outgroup, then we must not draw any conclusions regarding outgroup relationships from whatever resolution may result. This is simply because all such resolutions must derive from homoplasy and such evidence cannot form the basis for inferring cladistic relationships.

Ultimately, simultaneous, unconstrained analysis dispenses with the need to assign *a priori* polarity to characters altogether. Data are simply collected for all taxa, ingroup and outgroup alike, and analysed in a single matrix. The resulting cladogram is then rooted between the ingroup and outgroup and it is only at this point in the analysis that character polarity is determined (Nixon and Carpenter 1993). Under the transformational approach to cladistics, polarity is a property of a character that must be determined prior to cladogram construction. However, it can now be seen that polarity is actually something that is inferred *from* a cladogram. This perspective is part of the 'taxic approach' to cladistics (Eldredge 1979), in which only the distributions of characters among taxa are used to hypothesize group membership. All other properties of both characters and groups are derived from the resultant cladogram. To devotees of the transformational approach, such concepts are anathema. However, as stated earlier in this chapter, the majority of the computer algorithms used to estimate most parsimonious cladograms actually generate unrooted networks. No account is taken of any *a priori* polarity decisions. It is only when these networks are output as cladograms that they are rooted, and usually this is done by placing the root at the outgroup node. Thus, whether by conscious decision or not, most cladists use the taxic approach and carry out simultaneous, unconstrained analyses.

Several other techniques for rooting cladograms have been suggested, but all have major defects. The most frequently encountered alternative method uses an artificial taxon for which each character coded with the putative plesiomorphic state. However, two distinct concepts are sometimes conflated in this approach and we must be careful to distinguish between them. The first concept treats the artificial taxon as a 'hypothetical ancestor' or 'groundplan' (recently re-invented as 'compartmentalization' by Mishler 1994). However, as we have already seen, devising such a groundplan solely with reference to the ingroup is equivalent to estimating the conditions at the ingroup node, which may lead to a solution that is only locally optimal. Alternatively, if the algorithmic method of Maddison *et al.* (1984) is correctly applied, then the outgroup node with its optimized states can be interpreted as a composite, all-plesiomorphic outgroup. Such a taxon usually has all the characters coded as zero and is thus referred to as an 'all-zero outgroup'. But real outgroup taxa are necessary prerequisites to the formulation of such an all-plesiomorphic, hypothetical outgroup. If these real outgroup taxa display no character heterogeneity, then they are perforce equivalent to an artificial all-plesiomorphic outgroup. However, if there is heterogeneity among the outgroup taxa, the zero state cannot be assumed *a priori* to be the plesiomorphic state for the ingroup. Consequently, it is more efficient simply to code real outgroup taxa and employ them directly in a simultaneous, unconstrained analysis (Nixon and Carpenter 1996*a*).

When a set of heterogeneous outgroup taxa is used, it is frequently found that substitution of one of these taxa by another markedly alters the ingroup topology. Various strategies of outgroup selection to minimize such unwanted effects were discussed by Smith (1994*b*).

Another method that attempts to circumvent this perceived pernicious effect is Lundberg rooting. First, the most parsimonious network for the ingroup alone is determined, with no *a priori* assumptions regarding polarity. Then, keeping this topology fixed, an all-zero outgroup or hypothetical ancestor is attached to the branch that gives the least increase in length of the overall cladogram. However, Lundberg rooting suffers from all of the defects associated with all-zero artificial taxa discussed above. It has been suggested that a real outgroup should be used instead but the result would be most unlikely to be globally most parsimonious. Again, inclusion of the outgroup in a simultaneous, unconstrained analysis would be preferable (Nixon and Carpenter 1993).

In the unlikely scenario that there is no outgroup known that can be considered close enough for meaningful comparisons of characters to be made, and no ontogenetic information is available, then midpoint rooting has been suggested (Farris 1972). In this method, the root is placed at the midpoint of the longest path connecting two taxa in the network. Thus, besides the synapomorphies themselves, midpoint rooting also considers the amount of difference between taxa. Midpoint rooting can only be successful if the two most divergent taxa in the network have the same rates of evolutionary change, that is, a constant evolutionary clock must be assumed. This assumption, however, has both serious theoretical and empirical difficulties and thus midpoint rooting should be avoided.

De Pinna (1996) considered that rooting should use an optimality criterion, the only biologically defensible one of which is information pertaining to ontogenetic character transformation. Application of 'ontogenetic rooting' (de Pinna 1994) would thus yield a cladogram that was not only the most informative with regard to characters, but would also maximize the information available from observed ontogenetic transformations. We might indeed be fortunate enough to observe an ontogenetic transformation that would enable us to polarize the entire ingroup cladogram. However, it is much more likely, especially given our still fragmentary knowledge of most ontogenetic transformations, that ontogenetic rooting would allow us only to place an upper bound on the position of the root. This root, being most likely within the ingroup, would be considered by most systematists to be inferior to one placed between the outgroup and ingroup.

Thus, given that all other methods for rooting cladograms are either deficient or incomplete, we advocate simultaneous, unconstrained analysis, followed by rooting between the ingroup and outgroup, as the theoretically and empirically most defensible approach to cladogram construction (Nixon and Carpenter 1993).

3.3 CHAPTER SUMMARY

1. Exact methods of cladogram estimation guarantee discovery of the most parsimonious cladograms. However, they are time-consuming to implement and generally should not be considered for problems involving more than 25 taxa. The most widely implemented exact method is branch-and-bound.

2. For data sets of more than 25 taxa, heuristic (hill-climbing) methods are used, which sacrifice the certainty of finding the most parsimonious cladograms for computational speed. In order to improve the chances of finding the globally optimal solution, various algorithmic devices can be employed, including different addition sequences and branch swapping routines.

3. Of the various criteria proposed to determine character polarity, only the ontogenetic criterion and outgroup comparison have both a valid theoretical basis and wide applicability.

4. Cladistic analysis by means of the two-stage, constrained approach, which includes *a priori* determination of character polarity can lead to suboptimal results.

5. Simultaneous, unconstrained analysis, in which character polarity is derived from a cladogram, will never yield a less parsimonious cladogram than a two-step, constrained analysis and will often give a more parsimonious result. It is thus advocated as the preferred method of cladistic analysis.

4.
Optimization and the effects of missing values

4.1 OPTIMALITY CRITERIA AND CHARACTER OPTIMIZATION

Once the set of most parsimonious cladograms has been found, cladists generally wish to test hypotheses of character transformation. The first stage of this process is character optimization, which is effected by minimizing a quantity termed an optimality criterion. In Chapter 2, we introduced the concepts of additive (ordered) and non-additive (unordered) characters. These character types are equivalent, respectively, to the two basic optimality criteria, Wagner and Fitch optimization. However, there are innumerable other ways in which characters may be constrained to change. We now explain how the most commonly encountered optimality criteria are implemented to give what is termed a most parsimonious reconstruction (MPR) of character change. It should be noted, however, that the optimality criterion applied to each character is actually decided *prior* to cladogram construction. Furthermore, because each optimality criterion implies different costs, measured as number of steps, these choices will exert a large influence upon the length, and hence the topologies, of the most parsimonious cladograms for the data. Thus the reasons for choosing particular optimality criteria should be clearly explained and justified.

4.1.1 Wagner optimization

Wagner optimization, which was formalized by Farris (1970), is so named because it is based upon the work of Wagner (1961). It is one of the two simplest optimality criteria, imposing minimal constraints upon permitted character state changes. Free reversibility of characters is allowed. For binary characters, this means that a change from $0 \rightarrow 1$ is equally as probable as a change from $1 \rightarrow 0$. Similarly for a multistate character, a change from $1 \rightarrow 2$ is equally probable as a change from $2 \rightarrow 1$. However, in Wagner optimization, for a multistate character to change from 0 to 2, it is necessary to pass 'through' 1, and such a transformation will add two steps (i.e. $0 \rightarrow 1$, $1 \rightarrow 2$) to the length of the cladogram. Wagner optimization, therefore, deals with additive or ordered characters. A consequence of this reversibility is that the number of character state changes, and thus the length of the cladogram, is independent of the position of the root. An unrooted cladogram evaluated using Wagner optimization can be rooted at any point without changing its length.

In order to determine the minimum number of changes for a character using Wagner optimization, only a single pass through the cladogram is required, beginning with the terminal taxa and proceeding to the root. Consider an unrooted cladogram (Fig. 4.1a), in which six taxa (A–F) show four of five states of a multistate character (states 0, 1, 2, 4). First, one of the terminal taxa (A) is chosen arbitrarily as the root (Fig. 4.1b), although in practice, an outgroup taxon would usually fulfil this role. Optimization begins by choosing pairs of terminal taxa. The state(s) (termed the 'state set') assigned to the internal node that unites them is then calculated as the intersection of the state sets of the two derivative nodes. If the intersection is empty, then the smallest closed set that contains an element from each of the derivative state sets is assigned. For example, consider taxa E and F, linked by internal node z. The intersection of their state sets, (2) and (4) respectively, is indeed empty and thus the smallest closed set, (2–4), is assigned to z, and a value of 2 (i.e. $4 - 2$) is added to the cladogram length. Similarly, the intersection of the state sets of taxa C and D is also empty. Thus the state set (1–2) is assigned to their internal node y, with 1 being added to the length. In contrast, the intersection of the state sets of nodes y and z is not empty, as each contains the value 2. This value is assigned to their internal node x and no increment is made to the length. We proceed in this way towards the root until all internal nodes have been assigned state sets (Fig. 4.1c).

When this process is complete, the length of the cladogram will have been calculated, which for Fig. 4.1c is 5. However, it can be seen that this method does not necessarily assign states unambiguously to the internal nodes; that is, it does not produce a most parsimonious reconstruction (MPR). For example, we are uncertain whether node y should be assigned a value of 1 or 2. In order to produce an MPR, a second pass through the cladogram must be performed, this time starting from the root and visiting each internal node in turn. If the state set of an internal node is ambiguous, then we assign the state that is closest to the state found in the internal node of which it is a derivative. For example, nodes y and z are both assigned a value of 2 because this is the value in both their state sets that is closest to the value assigned to node x. Notice also that two changes ($2 \to 3$ and $3 \to 4$) must be assumed to have occurred between node z and taxon F. Once this process has been completed, then the MPR has been found and all five steps required by the character are accounted for (Fig. 4.1d).

It should be noted that this procedure (Farris 1970) will give a unique MPR only when all characters are free of homoplasy. In the presence of homoplasy, more than one MPR may exist. For example, Fig. 4.1e, in which nodes x and y are assigned state 1 rather than 2, also has five steps. However, there is a

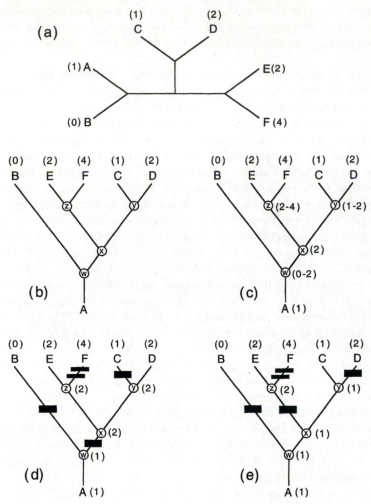

Fig. 4.1 Determination of character length using Wagner optimization (additive or ordered characters). (a) Unrooted cladogram for six taxa and considering one multistate character (states 0–4). (b) The unrooted cladogram is arbitrarily rooted at taxon A. (c) States assigned to internal nodes by passing from the terminal taxa to the root. (d) Alternative states for the internal nodes resolved by passing from the root to the terminals. (e) An alternative equally parsimonious resolution. The points at which character changes must be assumed are denoted by black bars. See text for details.

difference in the behaviour of the character changes. In Fig. 4.1d, the change from 1 → 2 was placed on to the cladogram at the closest possible position to the root, with the result that the occurrence of state 1 in taxon C must be accounted for by a reversal (2 → 1). This is known as 'accelerated' or 'fast' transformation (ACCTRAN in the PAUP program), because, when viewed from the root, 'forward' changes (e.g. $0 \to 1$ or $7 \to 8$) are placed on the cladogram as soon as possible. Accelerated transformation favours the acquisition of a character, with subsequent homoplasy accounted for by reversal. In contrast, Fig. 4.1e suggests that state 2 has been independently acquired twice, once between nodes x and z and once in taxon D. This is 'delayed' or 'slow' transformation (DELTRAN in the PAUP program), because it attempts to place forward changes on to a cladogram as far as possible from the root. Delayed transformation favours independent gains of a state rather than acquisition and reversal. It should be noted that under accelerated transformation (Fig. 4.1d), state 2 is still interpreted as an homology (albeit with subsequent reversal in taxon C), but under delayed transformation (Fig. 4.1e), state 2 in taxa E and F and state 2 in taxon D are considered to be two separate homologies and our original hypothesis of primary homology is refuted. The implications of and the choice between accelerated and delayed transformation are discussed further below (see §4.2).

4.1.2 Fitch optimization

Fitch optimization (Fitch 1971) works in the same way as Wagner optimization but concerns non-additively coded (unordered) characters. Once again free reversibility is allowed but for multistate characters there is equal cost, measured as steps, in transforming any one state into any other. Therefore, the changes $0 \to 1$, $0 \to 4$ and $2 \to 0$ all add a single step to the length of a cladogram.

Fitch optimization is implemented in a similar way to Wagner optimization but with two important differences. First, the state set assigned to the internal node is calculated as the *union* of the derivative state sets and the cladogram length is increased by one when the intersection of the two derivative state sets is empty. Second, when calculating an MPR, the state set assigned to an internal node x is that of the next more inclusive node, if this value is included within the state set of node x. Otherwise, any value in the state set of node x is chosen arbitrarily.

Fitch optimization is illustrated in Fig. 4.2, using the same initial cladogram as before (Fig. 4.2a). The cladogram is again arbitrarily rooted using taxon A (Fig. 4.2b) and the state sets assigned to the internal nodes during the first pass are z (2,4), y (1,2), x (2) and w (0,2) (Fig. 4.2c). The second pass up the cladogram then assigns the unambiguous state 2 to each of nodes w, y and z (Fig. 4.2d). Note that the assignment of state 2 to node w is arbitrary because the state found in taxon A (1) is not present in the state set for node w (0,2). Furthermore, because every state is only a single step from all other states, the branches linking node w to taxon B and node z to taxon F are each only one step long. Thus Fitch optimization produces an MPR that is only four steps long.

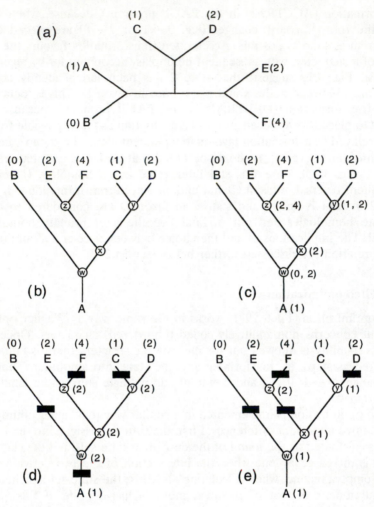

Fig. 4.2 Determination of character length using Fitch optimization (non-additive or unordered characters). (a) Unrooted cladogram for six taxa and considering one multistate character (states 0–4). (b) The unrooted cladogram is arbitrarily rooted at taxon A. (c) States assigned to internal nodes by passing from the terminal taxa to the root. (d) Alternative states for the internal nodes resolved by passing from the root to the terminals. (e) An alternative equally parsimonious resolution. The points at which character changes must be assumed are denoted by black bars. See text for details.

Fitch optimization, like Wagner optimization, does not necessarily produce a unique MPR. An alternative MPR is shown in Fig. 4.2e, where state 1 is assigned to node w. It is not readily apparent by examining the original state set (0,2) that state 1 is a possible unique assignment for node w. To discover all possible MPRs, a second pass through the cladogram is necessary.

In comparing these two basic methods of optimization, we would note that additive characters generally require more steps to be added to cladogram length than do non-additive characters, leading to more possible MPRs for the latter method.

4.1.3 Dollo optimization

In both Wagner and Fitch optimization, character states are allowed free reversibility. However, there are situations in which character states may be constrained in such a way that certain transformations are considered either highly unlikely or impossible. Dollo optimization was introduced in order to accommodate evolutionary scenarios in which it was considered most plausible *a priori* that each apomorphic state could only have arisen once and that all homoplasy must be accounted for by secondary loss. For example, in morphological studies, it may be thought that complex structures, such as the vertebrate eye, could only have arisen once. Similarly, in the molecular field, empirical studies have suggested that in analysis of restriction endonuclease cleavage map data for mitochondrial DNA, there is a marked asymmetry between the low probability of gaining a new site and the high probability of losing sites (DeBry and Slade 1985).

Dollo optimization requires that character polarity is prespecified. An example is shown in Fig. 4.3. In order to simplify the procedure, the cladogram is first (re-)rooted using one of the terminal taxa with the most derived state (Fig. 4.3b). State sets are assigned to the internal nodes and the length of the cladogram calculated as follows.

- If the state sets of two derivative nodes are equal, then this value is assigned to the internal node connecting them and cladogram length is not increased.

- If the state sets are different, then the *higher* value is assigned to the internal node, and cladogram length is increased by the difference between the two derived state sets.

- When the basal internal node is reached, its state set is compared with that of the root. If they differ, cladogram length is increased by the difference; otherwise no action is taken.

When applied to the cladogram in Fig. 4.3b, Dollo optimization produces an MPR of four steps (Fig. 4.3c). This exercise is performed using a root, but these rules have been modified to a generalized unrooted model (Swofford and Olsen 1990). Under this model it is not necessary to specify the derived state, but the polarity will be determined by the state at the root.

The disadvantage of Dollo optimization is that if the assumption regarding highly asymmetrical transformation costs is false, then levels of homoplasy and cladogram length will be severely overestimated. If we consider the

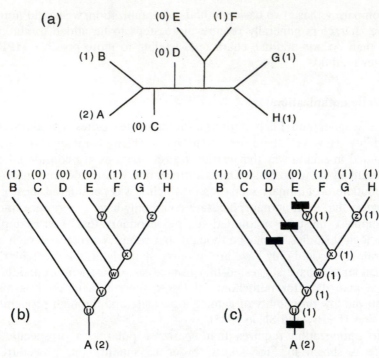

Fig. 4.3 Determination of character length under Dollo optimization. (a) unrooted cladogram for eight taxa and a character with three states (0, 1, 2). (b) The cladogram is rooted at taxon A (in this case this taxon shows the most derived state). (c) Assignment of character states to internal nodes. See text for details.

cladogram in Fig. 4.4a, Dollo optimization requires seven steps: a single origin of state 1 followed by 6 reversals to state 0. However, if the transformation costs are really equal, then Wagner or Fitch optimization ought to be applied, giving a cladogram on which only two convergent developments of

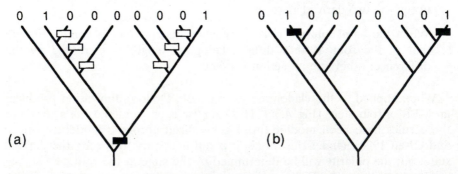

Fig. 4.4 Comparison of character length using Dollo and Fitch optimization. (a) A cladogram requiring seven steps using Dollo optimization (one gain and six reversals). (b) The same character requires only two steps using Fitch optimization.

state 1 need be postulated (Fig. 4.4b). Dollo optimization overestimates the length by five steps. The only means of avoiding this problem is to implement a 'relaxed' Dollo method, whereby one might prefer one gain and two losses to two independent gains, but reject one gain and ten losses in favour of two independent gains. A method for implementing such assumptions is discussed under 'generalized optimization' below.

4.1.4 Camin–Sokal optimization

Camin–Sokal optimization (Camin and Sokal 1965) constrains character transformations such that once a state has been acquired it may never be lost. Thus, any homoplasy must be accounted for by multiple origin. The method for calculating the number of steps on the cladogram (Fig. 4.5) is very similar

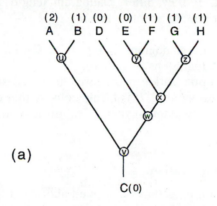

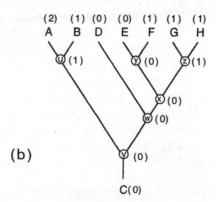

Fig. 4.5 Determination of character length under Camin–Sokal optimization. (a) cladogram of eight taxa and a character with three states (0, 1, 2). It must be rooted with a taxon showing the most plesiomorphic state. (b). States assigned to nodes after the rules mentioned in the text.

to that for Dollo optimization. Camin–Sokal optimization applies only to rooted cladograms in which the root bears the plesiomorphic state. If the root is not the plesiomorphic state, then character polarity must be reinterpreted to make it so. State sets are assigned to the internal nodes using the following rules.

- If the state sets of the two derivative nodes are equal, then this value is assigned to the connecting node and cladogram length is not increased.

- If the state sets are different, then the *lower* value is assigned to the connecting node and cladogram length is increased by the difference between the two derived states.

- When the basal internal node is reached, its state set is compared with that of the root. If they differ, cladogram length is increased by the difference. Otherwise no action is taken.

When applied to the cladogram in Fig. 4.5a, Camin–Sokal optimization produces an MPR of four steps (Fig. 4.5b).

Unlike the other optimization procedures described so far, Camin–Sokal optimization is very rarely used. It is highly unlikely that evolutionary scenarios would include the assumption that a feature may arise more than once but never be lost.

4.1.5 Generalized optimization

All of the optimization procedures described above can be treated as special cases of a generalized method of optimization. Under generalized optimization ('generalized parsimony' of Swofford and Olsen 1990), a 'cost' is assigned to each transformation between states (this concept was introduced briefly earlier in §2.4.6). The costs are represented as a square matrix, the elements of which represent the increase in cladogram length associated with the transformation of one state into another (Sankoff and Rousseau, 1975). Cost matrices for Wagner, Fitch, Dollo and Camin–Sokal optimizations are shown in Table 4.1. For Wagner optimization, it can be seen that the cost of transforming states through the series is cumulative, whereas for Fitch optimization, the cost of transforming between any two states is 1. In Dollo optimization, M represents an arbitrarily large number that guarantees a single forward transformation only on the cladogram. The infinite cost of reversals in the Camin–Sokal matrix prevents such transformations from occurring.

The advantage of generalized optimization is that it allows flexibility in permitted transformations that may not be otherwise available. For example, in nucleotide sequence data, transversions could be assigned different costs from transitions (see Chapter 5). Nor need the cost matrix be symmetrical. To

Table 4.1 Generalized parsimony. State × state cost matrices for four types of parsimony. Under the Dollo option, an arbitrary high value, M, is applied so that each gain of a character occurs only once on a cladogram. Under the Camin–Sokal option, reversals are prohibited by applying a value of infinity

	Wagner				Fitch				Dollo				Camin–Sokal			
	0	1	2	3	0	1	2	3	0	1	2	3	0	1	2	3
0	—	1	2	3	—	1	1	1	—	M	2M	3M	—	1	2	3
1	1	—	1	2	1	—	1	1	1	—	M	2M	∞	—	1	2
2	2	1	—	1	1	1	—	1	2	1	—	M	∞	∞	—	1
3	3	2	1	—	1	1	1	—	3	2	1	—	∞	∞	∞	—

implement 'relaxed' Dollo optimization, a suitable value of M is chosen so that the cost of a forward transformation is greater than that for a reversal but does not preclude multiple gains altogether. For example, if M is set to 1, then the upper triangle of the Dollo optimization cost matrix in Table 4.1 becomes the same as that for Wagner or Camin–Sokal optimization. Under this assumption, a single gain followed by multiple loss is the preferred hypothesis until the number of reversals exceeds four, after which two independent gains becomes the preferred hypothesis.

There are two difficulties in implementing generalized optimization procedures. The first is purely practical: the inclusion in a data set of characters coded using cost matrices greatly increases computation time. The second problem concerns the determination of the costs to be applied to transformations, which are dependent upon acceptance of a particular model of character change. Such models are mostly very difficult to defend *a priori* and as a general rule, unless the application of differential costs can be explicitly and thoroughly justified, then complex cost matrices should be avoided.

4.2 MISSING VALUES

Missing values, designated as '?', '-' or '*' in computer programs, are sometimes entered in data matrices. Most often, missing values appear in analyses containing fossil taxa and the problems that flow from their inclusion may be most acute in palaeontological data. However, missing data are not confined to fossils. There are a variety of circumstances in which question marks are used. This section explains the causes, effects and possible strategies for dealing with missing values.

Missing values may appear in a data matrix for one of several reasons.

1. A particular observation has not been scored even though the part of the animal or plant is available.

2. A question mark may be inserted in place of polymorphic coding (different members of a terminal taxon may show some or all of the alternative character states).

3. Technical difficulties in identifying a purine or pyrimidine in a gel may lead to the inclusion of a question mark or IUPAC/IUB ambiguity code within a molecular sequence, meaning that the observation is uncertain.

4. Once amino acid or gene sequences have been aligned between different species, it is often necessary to include insertions and deletions (indels) to maximize alignment of sequences. Usually indels are treated as an additional character state for the purpose of cladistic analysis. Occasionally, they are treated as question marks.

5. For fossils, which are always incomplete in one respect or another, there may be genuine missing data, that is, the particular part of the skeleton has not yet been found or the fossil lacks the soft anatomical details (fossils will nearly always lack molecular sequences).

6. Organisms may have missing data in the sense that some structures are interpreted as having been lost. For example, in a systematic problem concerning the interrelationships of vertebrates, attributes of tooth shape or tooth replacement pattern cannot be scored for modern turtles and modern birds.

7. Alternatively, there may be characters for which states cannot be coded in any outgroup taxon. For example, in the same systematic problem, states relating to features of the vertebral column cannot be scored in any outgroup (e.g. *Amphioxus* or sea squirts) because no outgroup has a vertebral column. In both this and the previous case, character states may be judged non-applicable or illogical for some taxa, and although this is usually noted in the written data matrix, computer algorithms will treat these entries as question marks.

8. A special case is exemplified by three-item statements analysis, where data are recoded to express components. Here, the question marks are simply devices, which the computer algorithm can accept. They do not indicate missing or ambiguous character data. This type of question mark is discussed further in Chapter 9.

Irrespective of the cause of question marks in a data matrix, their effect in a computer parsimony analysis is the same. Question marks can lead to the generation of multiple equally most parsimonious cladograms, to spurious theories of character evolution, and to lack of resolution by masking the phylogenetic signal implied by the observed data.

It is important to recognize that missing values alone will not alter the topological relationship of taxa that are known from real (observed) data.

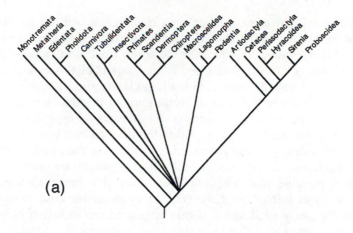

(a)

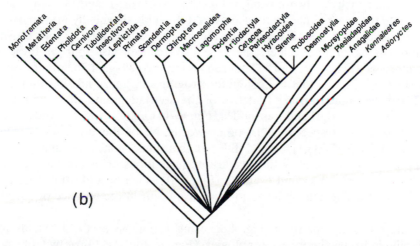

(b)

Fig. 4.6 An example of the increase in the number of equally most parsimonious cladograms following addition of taxa with missing values. Above (a) is the strict consensus tree for 20 orders of Recent mammals coded for 88 characters with nearly all data cells filled with real values. Below (b), seven fossil taxa have been added, each of which have between 25–57% missing data. (From Novacek 1992.)

Their most obvious effect is to increase the number of equally most parsimonious cladograms and decrease resolution. Novacek (1992) gave a good example of this phenomenon (Fig. 4.6). An initial analysis of the twenty recognized orders of living mammals using 88 morphological characters produced four equally most parsimonious cladograms. All but six data cells were filled with real data (i.e. 0 or 1), the few question marks being the result of non-applicable coding. To this analysis, seven fossil taxa were added with varying amounts of missing data (25–57%). Analysis of this enlarged matrix resulted in 6 800 + equally most parsimonious cladograms (this being the limit of computer memory rather than the total number of equally most parsimonious solutions). In the strict consensus tree, the clade originally recognized as containing primates, tree shrews, flying lemurs and bats was lost.

Addition of taxa to any analysis is liable to increase the number of equally most parsimonious cladograms by introducing additional homoplasy (although it is possible that additional taxa may give fewer cladograms by resolving previous ambiguity). However, the dramatic increase in cladogram number in the study cited above is due largely to the inclusion of question marks, which increase ambiguous character optimizations at the internal nodes. It has been pointed out by Platnick *et al.* (1991*b*) that two of the most commonly used parsimony programs (Hennig86 and PAUP) can generate spurious cladograms when supplied with matrices containing missing data. The example of Platnick *et al.*, reproduced here as Fig. 4.7, shows a data matrix that includes two missing values for two taxa (F and G). Analysis of this data set using either Hennig86 or PAUP yields six equally most parsimonious cladograms. Yet, if we replace the missing values with all four possible combinations of real values, then we can recover just four of these six cladograms (Fig. 4.7a, d–f). The remaining two (Fig. 4.7b, c) are solely products of the way that the computer programs optimize missing values.

It can be argued further that five of the six cladograms, those with three nodes (Fig. 4.7a, d, f) and four nodes (Fig. 4.7b, c), are 'over-resolved', by one and two nodes respectively. In other words, all of the branches resolving groups DEFG, DFG, DG and DF are spurious; none has unambiguous support in the data. How these spurious resolutions arise can be explained with reference to a second example (Fig. 4.8), which also illustrates that the problem of ambiguous optimization can also occur in the absence of question marks.

Analysis of the matrix in Fig. 4.8a using Hennig86 or PAUP yields four equally most parsimonious topologies (Fig. 4.8b–d, f). Only two of these cladograms are unambiguously supported by data: Fig. 4.8b, in which node CDE is supported by character 3; and Fig. 4.8c, in which node BC is supported by character 4. The cladograms shown in Fig. 4.8d and e, and in Fig. 4.8f and g, represent alternative, ambiguous optimizations of characters 4 and 3 respectively. The former cladogram of each pair is the delayed transformation, while the latter cladograms are the accelerated transformations. The topologies are reported by the computer programs because at least one character is placed on each branch under at least one of these optimizations (Fig. 4.8e and Fig. 4.8f).

Swofford and Begle (1993) proposed three criteria that could be used to determine when zero-length branches are to be collapsed.

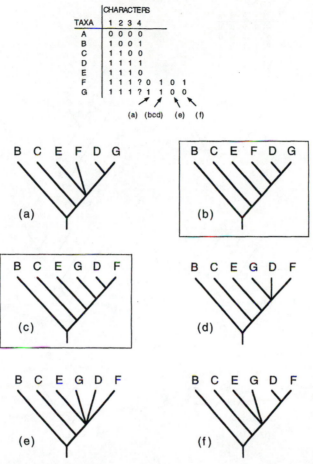

Fig. 4.7 Spurious cladograms. (a–f) When the data set shown here, including two questions marks, is analysed using either Hennig86 or PAUP, six cladograms (a–f) are found. However, two of these (b and c) cannot be justified by any combination of replaced 'real' observations. they are spurious and result from the way the algorithms treat zero branch lengths. (From Platnick *et al.* 1991b.)

- Collapse a branch if its *minimum* length is zero; that is, if there is at least one optimization of all characters that assigns zero-length to the branch, then that branch is collapsed.

- Collapse a branch if its *maximum* length is zero; that is, if there is at least one optimization of all characters that does not assign zero-length to the branch, then that branch is not collapsed.

- Apply either accelerated or delayed transformation to all characters or to characters individually, then collapse any branch assigned zero-length.

The third criterion appears arbitrary, in that rarely would we have a

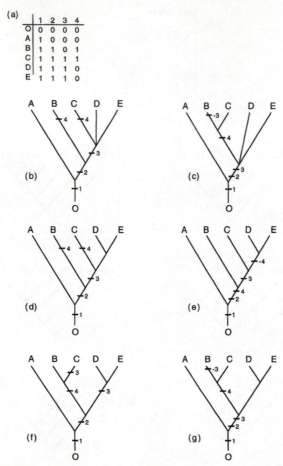

Fig. 4.8 (a–g) Example of how ambiguous optimization can lead to spurious resolution when a data set is analysed using either PAUP or Hennig86; see text for explanation.

defensible justification for choosing one type of optimization over the other. However, de Pinna (1991) argued that, unless demonstrated to be false by parsimony, accelerated transformation is to be preferred because it preserves more of our original conjecture of primary homology than does delayed transformation. (The same point was made by Farris (1983) on the basis of higher information content.) In other words, by favouring the acquisition of a character, with subsequent homoplasy accounted for by reversal (e.g. Fig. 4.8e rather than Fig. 4.8d), accelerated transformation maintains our original conjecture of the character as a putative synapomorphy. In contrast, by treating homoplastic characters as independent derivations, delayed transformation rejects our original hypothesis of primary homology. For this reason, de Pinna (1991) asserted that accelerated transformation optimization is the theoretically superior algorithm for tracing character evolution. It should be noted, however, that the argument of de Pinna explicitly concerns character *evolution*. It is therefore part of the transformational approach to cladistics, which views characters as features that transform one into another, rather than the taxic approach, which we adopt here, that uses the distributions of characters among taxa to hypothesize group membership.

Swofford and Begle (1993) considered that the first criterion was flawed because it might not be possible to collapse all branches that have a minimum length of zero and still retain a most parsimonious topology. Two branches may each be potentially of zero-length, but not simultaneously. Thus, it is not possible to collapse both without increasing the cladogram length, giving a suboptimal solution. They considered that only the second criterion could be justified, maintaining that ambiguous support represented potential resolution and thus should be retained on the most parsimonious cladograms. Consequently, this approach was adopted in PAUP (and also Hennig86; Farris 1988).

However, there is a practical problem. PAUP forces us to choose to optimize using either accelerated or delayed transformation. Accelerated transformation should recover the topologies in Fig. 4.8b, c and e, but not that in Fig. 4.8g, which ought to collapse into the topology in Fig. 4.8c due to the lack of support for node DE. Likewise, delayed transformation should recover the topologies in Fig. 4.8b, c and f, but not that in Fig. 4.8d, where the lack of support for node DE ought to collapse it into the topology in Fig. 4.8b. Nevertheless, PAUP continues to report all four cladograms, even though one of them cannot be supported by data under the chosen optimization routine.

Coddington and Scharff (1994) suggested filtering cladograms produced using the third criterion by discarding all topologies that must contain a zero-length branch. This procedure would resolve the problem just discussed, as well as removing those cladograms in which not all nodes are capable of support simultaneously. However, it would retain topologies such as those in Fig. 4.8e and Fig. 4.8f, where the resolution is supported by one optimization, although not by another.

However, Nixon and Carpenter (1996b) noted that the extra resolution comes only from a small number of ambiguously distributed, homoplastic characters. The additional groups are thus weakly supported and are not strong hypotheses of relationship. Instead, they suggested that the most efficient way to summarize the grouping information in the data is to eliminate all those minimum length cladograms that contain spuriously re-solved nodes. In other words, apply the first criterion of Swofford and Begle, but with the added proviso that the minimum cladogram length be main-tained, thereby circumventing Swofford and Begle's objection. Those clado-grams that remain, which are both of minimum length and have all the resolved nodes supported by data, are termed 'strictly supported cladograms' (Nixon and Carpenter 1996*b*) (the related concept of 'minimal cladograms' (Nelson 1992) in the context of three-item statements analysis is discussed in Chapter 9). In the example in Fig. 4.8, only the two cladograms in Fig. 4.8b and Fig. 4.8c would be retained. Of the programs currently available, only NONA has a switch to disallow ambiguous optimizations.

To return now to missing data, the effects can be quite subtle because there is no simple relationship between the amount of missing data and the disruptive influence it may exert on either the number of cladograms produced or their resolution. This is because the effect of introducing a taxon with many question marks depends upon the distribution of the remaining real (observed) data. This can be illustrated by an example published by Nixon and Wheeler (1992) and redrawn here as Fig. 4.9. Given six taxa, A–F, each of which is completely known by real data, there is a single most parsimonious solution (Fig. 4.9a). The addition of a seventh taxon, G, which is known only for characters 3 and 6, results in eight equally most parsimonious solutions, the strict consensus of which is the unresolved bush (Fig. 4.9b). This is because the real data known for taxon G places it in markedly different parts of the original cladogram (Fig. 4.9c). Therefore, in a real analysis, if there is a taxon (e.g. a poorly known fossil) that behaves in this manner, it may be better to leave it out entirely. Instead, it could be placed by hand on the resulting cladogram in the alternative positions allowed by the real data.

The removal of a taxon simply because it disrupts resolution cannot be justified except in terms of computational expediency. A taxon, with or without question marks, that causes a change in topology, other than simply decreasing resolution, carries information that may be potentially useful. Any taxon that has a unique combination of characters may influence relationships among the other taxa. It is against this background that Wilkinson and Benton (1995) suggested employing 'safe taxonomic reduction'. This procedure permits us to eliminate those taxa with question marks that can have no influence on the topology of the remaining taxa. A theoretical example is shown in Fig. 4.10. Given the three taxa (A, B, C), it can be seen that taxon B has precisely the same real codings as the more completely known taxon A. Deletion of this taxon cannot alter topological relationships, yet its presence may increase the number of equally most parsimonious cladograms. Taxon C contains exactly the same number of question marks as taxon B but they are distributed differently such that, for character number 2, taxa A and C have different real codings. Elimination of taxon C would be unsafe because the real data for this taxon may have influence on the resulting topology. In a real example, Wilkinson and Benton (1995) analysed 16 taxa of Rhynchosauridae (an extinct family related to the tuatara), finding 21 700 equally most parsimonious cladograms (the limit imposed by the computer's memory). Six taxa satisfied the criterion of safe taxonomic reduction and could be

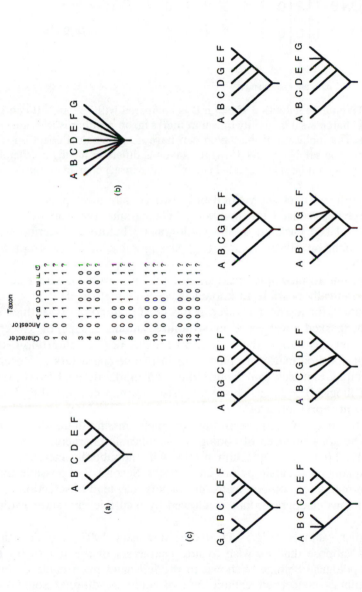

Fig. 4.9 Some taxa with question marks can be highly disruptive. (a) The single cladogram that results from analysis of the data set shown, but including only taxa A–F. (b) When a seventh taxon, G, is added to the analysis, eight cladograms are produced, of which the strict consensus tree is the uninformative bush. (c) The disruption is caused by the alternative positions that taxon G can adopt on the eight original cladograms. (From Nixon and Wheeler 1992.)

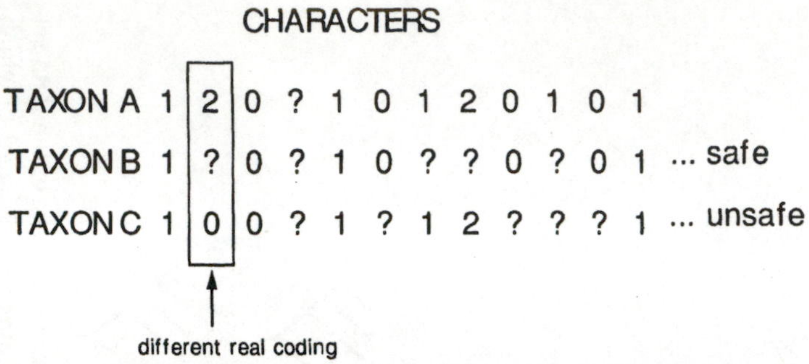

CHARACTERS

TAXON A 1 | 2 | 0 ? 1 0 1 2 0 1 0 1
TAXON B 1 | ? | 0 ? 1 0 ? ? 0 ? 0 1 ... safe
TAXON C 1 | 0 | 0 ? 1 ? 1 2 ? ? ? 1 ... unsafe

↑
different real coding

Fig. 4.10 Safe taxonomic reduction. If taxon B is compared with taxon A, it can be seen that in all characters denoted by question marks taxon A either is the same or has real values. The inclusion of this taxon can have no topological effect on the outcome and may be safely deleted. However, taxon C differs in having a different 'real' value and may not be safely deleted since it may cause a change in topology.

deleted. The reduced data set yielded only two equally most parsimonious cladograms. Having reduced the number of cladograms, the excluded taxa would then be placed back on to the cladogram(s) before a selection was made among them on other grounds (e.g. stratigraphic or biogeographical plausibility).

Missing data due to incompleteness should not be a problem with Recent taxa. It is theoretically possible to know a Recent animal or plant in a way that is not possible for fossils. However, it has been pointed out (Gauthier *et al.* 1988) that marked divergence in structure among groups of Recent animals and plants, as well as the inclusion of highly distinct outgroup taxa, may mean that question marks are placed against some characters in Recent taxa. For example, matrices can include question marks derived from non-applicable (or illogical) codings. This often arises in matrices including both fossil and Recent representatives.

Although little may be done regarding genuinely missing data, we should be aware of the consequences of coding non-applicable character states as question marks. First, these question marks will undoubtedly increase the numbers of equally most parsimonious cladograms. Second, it is possible that using question marks for non-applicable characters may lead to selection of a more parsimonious cladogram than is allowed by plausible character evolution.

The following example (Fig. 4.11, from Maddison 1993) explains this phenomenon. Suppose that we wish to add characters of the tail to try to resolve the topological ambiguity shown in the left-hand main clade of Fig. 4.11a. The animals concerned exhibit one of three conditions. Some are tailless, some have red tails and some have blue tails. The distribution of these conditions in our initial analysis is shown in Fig. 4.11a, where it can be seen that the more basal members of both the two major clades are all tailless, while the more distal members of each show either red tails or blue tails.

There are several different ways in which we might code the tail conditions (see Chapter 2). Two of the commonly used methods are shown here as alternative 1 and alternative 2 (Fig. 4.11b). Alternative 2 treats the three types as dependent variables within a single multistate character. In contrast, alternative 1 treats them as semi-independent variables distributed between two characters and includes a question mark to denote non-applicable coding for colour in those taxa having no tails. Coding alternative 1 is often used, yet this may lead to a selection of a cladogram that cannot possibly be justified in terms of character evolution (that is, when we translate the cladogram into a tree). The area of ambiguity in the left-hand clade involves four taxa, for which there are 15 possible fully resolved solutions. Two are shown in Fig. 4.11c and e, and in Fig. 4.11d and f. Optimization of the tail characters (presence/absence, red/blue) coded as alternative 1 results in us preferring the topology in Fig. 4.11c over the alternative (Fig. 4.11d), because the former requires only two steps to explain the distribution of tail colour, rather than three. But selection of that cladogram is nonsensical, because the choice is based upon falsely ascribing tail colour (blue or red) to animals that do not have tails. However, if we used coding alternative 2, the starting condition for both topologies would be '0' and both require 4 steps (Fig. 4.11e and Fig. 4.11f). Of course, coding tail attributes in this way gets us no further in resolving the initial polytomy but it does mean that we avoid the choice of an apparently optimal cladogram based on nonsensical character attributes. As Maddison pointed out, this problem is not confined to morphological data but is also relevant to the use of question marks to code for gaps in protein or nucleotide sequences. This is a very simple example and if we were doing the analysis by hand, then the danger would be spotted. But in computer analyses, these pitfalls are much more difficult to detect. Furthermore, if we were to apply successive approximations character weighting (see Chapter 5) to such an initially 'nonsensical' cladogram, the result may well deviate further from reality.

Simulations have shown that question marks in ingroup taxa that are widely scattered at low hierarchical levels exert more deleterious effects on character optimization (and perhaps false selection of cladograms) than do question marks in taxa near the root. In practical terms, this means that in combined analyses of fossil and Recent taxa, where the latter are usually scattered within the ingroup, it may be particularly important to avoid the use of question marks for non-applicable character states.

Mention has been made above of the use of question marks for polymorphic taxa. To avoid this two solutions have been proposed. First, the group may be broken up into two or more subgroups, the members of which show uniform coding. Alternatively, a phylogeny for the group showing the polymorphism may be assumed and the coding at the ingroup node accepted as

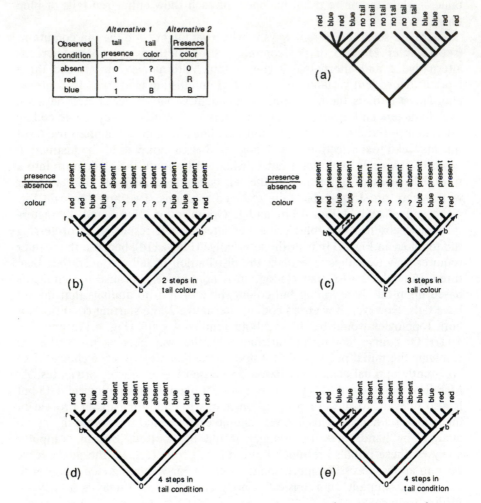

Fig. 4.11 An example of how non-applicable coding, scored as question marks, may result in choosing a cladogram that is 'nonsensical'. (a) The preliminary cladogram, where part of the left hand side of the cladogram was unresolved. We wish to add characters of the tail to try to resolve this polytomy. There are three taxon types (tailless, red-tailed and blue-tailed). (b) Two coding alternatives for the tail attributes. Alternative 1 treats the features as two characters and includes a question mark to denote inapplicable coding for colour in those taxa having no tails. Alternative 2 treats the three types as a single multistate character. (c) One of the fifteen possible resolutions of the terminal polytomy shown in (a). (d) Another of the fifteen possible resolutions of the terminal polytomy shown in (a). Optimization of the tail characters coded according to alternative 1 leads us to prefer the topology in (c) to that in (d), because the former is one step shorter. (e–f) The same two topologies as in (c) and (d). Optimization of the tail characters coded according to alternative 2 does not allow us to prefer either topology. Both are four steps long.
(From Maddison 1993.)

representative of the entire group (but see §3.2.5 for a discussion of the problems associated with this procedure).

In summary, the introduction of question marks into cladistic analyses causes computational problems that have not yet been solved. These problems relate to both the numbers of cladograms produced and their resolution. The introduction of question marks is clearly most significant when undertaking simultaneous analyses of combined fossil and Recent taxa (especially where molecular sequences are included). For morphological data matrices, it may be possible to distinguish between informative and non-informative fossil taxa and eliminate the latter. It may also be possible to identify 'rogue' taxa and eliminate them from initial analyses. For polymorphisms, alignment gaps in molecular sequences and non-applicable character states, care should be taken in the initial coding. All of these strategies will alleviate the symptoms of question marks but not remove them entirely.

4.3 CHAPTER SUMMARY

1. Character optimization is the process of determining the sequence of character state change on a cladogram. Wagner optimization is used for ordered characters and Fitch optimization for unordered characters. Other, more restrictive, methods include Dollo and Camin–Sokal optimization. All of these procedures can be considered to be special cases of a generalized method of optimization in which explicit costs are assigned to transformations between character states.

2. Missing values in a data matrix can be caused by the failure to observe a feature in a particular organism due to lack of the appropriate organ or life cycle stage, polymorphism, secondary loss, or dependent characters leading to non-applicable character states.

3. Missing values can lead to an increase in the number of equally most parsimonious cladograms, may decrease resolution, and may lead to selection of a more parsimonious cladogram than is allowed by plausible character evolution.

4. Missing values can also cause some cladistic computer programs to produce spurious cladograms that cannot be supported by any possible combination of real values. Such cladograms can also be produced from data sets that include no missing values as a result of ambiguous character optimization. The preferred cladogram is that which is of minimum length and has all its nodes unambiguously supported by data. This is the strictly supported cladogram.

5.
Measures of character fit and character weighting

5.1 MEASURES OF CHARACTER FIT

Current parsimony programs utilize a number of different statistics to assess the 'quality' of cladograms. Standard measures are cladogram length, the consistency index and the retention index. The examples given below deal with binary characters only. The principles can easily be extended to multistate characters.

5.1.1 Cladogram length

Consider Matrix 1 (Table 5.1a), which consists of four taxa (A–D) and six characters (1–6, in which 0 = the plesiomorphic state and 1 = the apomorphic state). With the inclusion of an all-zero root (X), this matrix yields three equally most parsimonious cladograms (Fig. 5.1a–c). Consider first the cladogram in Fig. 5.1a. Four of the six characters fit the cladogram with one step (Table 5.1, row s; characters 1, 4, 5 and 6). Character 1 is present only in taxon D and character 4 is present in all four taxa, A–D. No matter which solution is considered, characters 1 and 4 will always fit with one step. Characters 2, 3, 5 and 6 are different in that they imply alternative groupings among taxa A–D. Character 5 implies that A and B are more closely related to each than either are to C or D, while character 6 implies that A, B and C are more closely related to each other than they are to D. Both of these hypotheses are represented in Fig. 5.1a. Hence, characters 5 and 6 need only appear once on the cladogram at the node uniting the relevant taxa. Characters 2 and 3 are of the same kind as character 5 and 6 in that they also imply particular groupings. Character 2 implies that C and D are more closely related to each other than either is to A or B, while character 3 implies that B, C and D are more closely related to each other than they are to A. Neither of these hypotheses is present in Fig. 5.1a, on which we will now concentrate. Characters 2 and 3 can only be fitted to that cladogram with more than one appearance. In this case, characters 2 and 3 appear with a minimum of two steps (Matrix 1, Table 5.1a, row s). In total, in Fig. 5.1a, four characters fit the cladogram once (characters 1, 4, 5 and 6), and two characters fit the cladogram twice (characters 2 and 3). Simple addition gives a total cladogram length total of 8 steps.

Table 5.1 (a) Matrix 1 with four taxa (A–D) and an all-zero root (X) coded for 6 characters (1–6). (b) Matrix 2 with five taxa (A–E) and an all-zero root (X) coded for 6 characters (1–6). s = actual steps on the cladogram, m = minimum possible steps, g = minimum steps on a bush, ci = consistency index, ri = retention index, S = sum of s, m, and g values (S, M and G respectively) for the calculation of CI and RI. CI(u) is CI for informative characters only. The values for CI, CI(u) and RI for Matrix 1 refer to the cladogram in Fig. 5.1c, while those for Matrix 2 refer to the cladogram in Fig. 5.2c.

(a) Matrix 1

T			Characters				
	1	2	3	4	5	6	Σ
X	0	0	0	0	0	0	
A	0	0	0	1	1	1	
B	0	0	1	1	1	1	
C	0	1	1	1	0	1	
D	1	1	1	1	0	0	
s	1	2	2	1	1	1	8
m	1	1	1	1	1	1	6
g	1	2	2	1	2	2	10
ci	1	1	1	1	0.5	0.5	
ri	1	1	1	1	0	0	
Length				8			
CI		M/S	6/8	0.75			
CI(u)		M/S	4/6	0.67			
RI		G-S/G-M	10−8/10−6	0.50			

(b) Matrix 2

T			Characters				
	1	2	3	4	5	6	Σ
X	0	0	0	0	0	0	
A	0	0	0	1	1	1	
B	0	0	1	1	1	1	
C	0	1	1	1	0	1	
D	1	1	1	1	0	0	
E	1	1	1	1	0	0	
s	1	2	2	1	1	1	8
m	1	1	1	1	1	1	6
g	2	3	2	1	2	3	13
ci	1	1	1	1	0.5	0.5	
ri	1	1	1	1	0	0.5	
Length				8			
CI		M/S	6/8	0.75			
CI(u)		M/S	4/6	0.67			
RI		G-S/G-M	13−8/13−6	0.71			

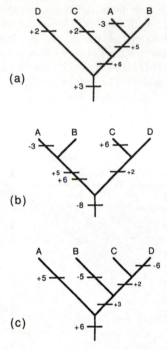

Fig. 5.1 a–c. Analysis of the data in Table 5.1a yields three equally most parsimonious cladograms. Only characters 2, 3, 5 and 6 are mapped. Character gain signified by '+', character loss by '−'.

It is tempting to consider the fit of a character and the terms 'steps', 'appearances' or 'occurrences' as if they represent real events, such as the origin of a particular character. The term 'steps', for example, has been interpreted either as the number of occurrences of a particular character on a particular cladogram, or as the amount of change required to transform a character from one of its states into another (in this case 0 to 1, or 1 to 0). Under this latter interpretation, the number of steps between different states of a character can be made to vary according to pre-specified 'optimality criteria' (see Chapter 4). Such assumptions can be incorporated into the data set *a priori* and considered as a form of weighting, such that one particular 'transformation' is considered more 'likely' than another.

Various assumptions to weight step changes differentially have been proposed. Wagner, Fitch, Dollo and Camin-Sokal 'parsimony' are among those more commonly encountered. The recently developed concept of 'generalized parsimony' attempts to incorporate even more complex changes into character transformations (see Chapter 4). However, such assumptions need not (and possibly should not) be made and cladogram length is simply a way of quantifying the optimal cladogram, given a particular data set and possible solutions.

When a character fits a cladogram perfectly (for a binary character, to one node with one step), it can be considered 100% consistent with that particular cladogram. When a character does not fit a cladogram perfectly (to more than one node with more than one step), it is less than 100% consistent. Such a character exhibits homoplasy. Homoplasy is discordance with a particular cladogram and can be measured by various indices: the consistency index (ci) and the retention index (ri) for individual characters, and their ensemble values (CI) and (RI) measured over the entire suite of characters for a particular cladogram(s).

5.1.2 Consistency index (ci)

Character consistency (ci) is defined as m/s, where s is the minimum number of steps a character exhibits on a particular cladogram and m is the minimum number of steps the same character can show on any cladogram. For a binary character on any cladogram, m will equal 1 (Matrix 1, Table 5.1, row m). The best possible fit will be to a single node and for such a character, s will also equal 1. However, with increasing amounts of homoplasy, a binary character will fit with a greater number of steps and s will equal 2 or more.

Consider again Matrix 1 (Table 5.1a) and the cladogram in Fig. 5.1c. Disregarding characters 1 and 4 as uninformative, the characters can be divided into those that fit to one node (characters 2 and 3) and those that fit to more than one node (characters 5 and 6). All four are binary characters and thus $m = 1$ for all of them. As characters 2 and 3 fit to a single node on the cladogram in Fig. 5.1a, their s also equals 1, and thus their ci $= 1$ ($m/s = 1/1$). They are 100% consistent with this cladogram. In contrast, characters 5 and 6 fit to the cladogram twice. Their $s = 2$ and hence their ci $= 0.5$ ($m/s = 1/2$). They fit this particular cladogram imperfectly and are only 50% consistent. It should be noted that for Matrix 1 in Table 5.1a, there are three equally parsimonious solutions (Fig. 5.1a–c) and the values of s (and hence ci) for characters 2, 3, 5 and 6 will change according to the cladogram being considered.

5.1.3 Ensemble consistency index (CI)

Values of ci for individual characters are useful to understand how well they have performed on different cladograms, but it is also useful to know how well the entire data set performs. The ensemble consistency index (CI) gives this value. As with individual character ci values, if all the data fit a cladogram perfectly then data and cladogram are 100% consistent. Data that are 100% consistent are rarely, if ever, encountered, no matter how well the morphology is studied or the DNA sequencing is performed. CI is represented by the same parameters as individual character consistency but with upper case letters to indicate their summation. Hence $CI = M/S$, where M

equals the sum of all the m values for the individual characters and S is the sum of all their s values. In the case of Matrix 1 (Table 5.1a), $M = 6$ (addition of values in row m) and $S = 8$ (addition of values in row s). Therefore, $CI = M/S = 6/8 = 0.75$. The data are 75% consistent with the cladogram.

5.1.4 Problems with the consistency index

Several problems have been noted regarding the use of the consistency index as a measure of homoplasy. First, the inclusion of uninformative characters in the calculation will inflate CI values. Consider once again the example of Matrix 1 (Table 5.1a). As noted above, in this particular data set two characters (1 and 4) are uninformative of relationships. Whatever possible solution is considered, their ci value will always equal 1. For this reason, their values might be understood as irrelevant or unnecessary (Yeates 1992). If they are deleted from consideration, the CI (M/S) becomes $4/6 = 0.67$. This figure is somewhat lower than the 0.75 that was obtained when the uninformative characters were included in the calculation. However, the significance of the two different values is probably only relevant when different data sets are being compared. If different parsimony analyses are performed with the uninformative characters included, then the values for various solutions will be affected by this (a position contested by Bryant 1995). Most currently available parsimony programs (e.g. NONA, PAUP) provide both values as part of the output.

Second, as the number of taxa increases, values of CI are observed to decrease. In most cases, when the number of taxa increases, the CI will decrease irrespective of any change in information content. However, this is a recognized and expected property. The intention behind CI is to measure the amount of homoplasy in a given character or data set. This it does adequately. Farris (1989) recognized the problem of character fit (synapomorphy), as opposed to homoplasy, and introduced the retention index, ri (and its associated ensemble value, RI), to account for this (see below).

The third perceived problem with the consistency index is that its value can never reach zero. All binary characters will have a best and worst value. In Matrix 1 (Table 5.1a), character 5, for example, will have a best value ci = 1 (1 step, 100% consistent) and a worst value ci = 0.5 (2 steps, 50% consistent). As the examples in Matrix 1 (Table 5.1a) and the cladograms in Fig. 5.1 deal with only 4 taxa, a problem does not arise. However, consider a binary character that is apomorphic for taxa ABC from a set of five, A–E. For the character to be 100% consistent it would need to group A, B and C together at a single node. If it performs less well, it may group only A + B (or A + C, or B + C) together at one node and would then have a ci of 0.5, i.e. one step for the group A + B and one step for C. If the character performs at its poorest it would not group any taxa together and A, B and C would appear on separate branches of the cladogram. In this case, ci = 0.33 (3 steps, or 33% consistent).

Thus it can be seen that binary characters will always have a positive c_i regardless of their relative informativeness; c_i can never reach zero. At this stage, it is worth remembering that c_i is intended to measure amount of homoplasy and is still a useful measure for that purpose. However, in situations such as the above example of a binary character that is apomorphic in three taxa from a set of five, there will be cases in which some (but not all) of the similarity can be interpreted as synapomorphy, such as when only A + B are grouped to the exclusion of C. To measure this, Farris (1989) proposed the retention index (r_i).

5.1.5 Retention index (r_i)

The retention index (r_i) is defined as $(g - s)/(g - m)$, where m and s are the same as for c_i, and g is the greatest number of steps a character can have on any cladogram ('any cladogram' can be understood to be the unresolved bush). Thus for binary characters, g will equal the fewest number of variables in the column. For example, in Matrix 1 (Table 5.1a), character 2 has two '1' values and three '0' values, hence $g = 2$; character 3 has three '1' values and two '0' values, hence $g = 2$. For binary characters that fit to one node of a cladogram, such as characters 2 and 3 in Fig. 5.1c, $r_i = (2 - 1)/(2 - 1) = 1$. For binary characters that do not fit to any node of a particular cladogram, as characters 5 and 6 in Fig. 5.1c, $r_i = (2 - 2)/(2 - 1) = 0$.

To appreciate the difference between what c_i and r_i measure, consider Matrix 2 (Table 5.1b). Matrix 2 is identical to Matrix 1 except for the addition of taxon E, which has exactly the same characters as taxon D. Three new cladograms result (Fig. 5.2a–c). The total cladogram length and the individual character c_i values are the same respectively as for the cladograms without taxon E (Fig. 5.1a–c), but the r_i values differ. For character 6 in Fig. 5.2c, $r_i = (3 - 2)/(3 - 1) = 0.5$, rather than 0 as in Matrix 1 (Table 5.1a, Fig. 5.1c), because the character now groups D + E together. Character 6 shows the 0 state in D + E, which is interpreted as synapomorphic. (The situation is the same with the cladogram in Fig. 5.2b.) Thus, although the amount of homoplasy for character 6 remains unchanged ($c_i = 0.5$), the evidential value of this character differs depending on the cladogram ($r_i = 0$ or 0.5).

Consider a further example illustrated in Fig. 5.3 (after Goloboff 1991). Character 1 is present in taxa D and H and can fit a cladogram in two possible ways: either the group D + H is supported ($c_i = 1$, $r_i = 1$) or it is not ($c_i = 0.5$, $r_i = 0$). On the cladogram in Fig. 5.3, character 1 fits the cladogram as poorly as it possibly could. It has a c_i of 0.5 and because none of the similarity is interpretable as synapomorphy, it has a r_i of 0. Character 2 is present in taxa A, B, C, F and I, allowing a number of possible groups to be supported (ABCF + I, ABC + FI, ABF + CI, and so on, up to the entire group ABCFI). There is a reasonable possibility that a proportion of the similarity among the taxa A, B, C, F and I will be interpreted as synapomorphy. Character 2 fits the cladogram in Fig. 5.3 poorly (3 times) and has a c_i of

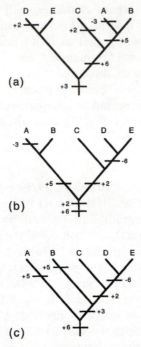

(a)

(b)

(c)

Fig. 5.2 a–c. Analysis of the data in Table 5.1b yields three equally most parsimonious cladograms. Only characters 2, 3, 5 and 6 are mapped. Character gain signified by '+', character loss by '−'.

0.33. However, it could have a worse fit as there are five taxa with the apomorphic value and hence its poorest ci would be $1/5 = 0.2$. For the cladogram in Fig. 5.3, character 2 does group some taxa together (A + B + C). Hence some of the similarity is interpretable as synapomorphy and its ri = 0.50.

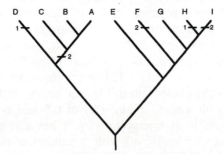

Fig. 5.3 Cladogram for nine taxa, A–I, with the two characters in Table 5.2 mapped. Character 1 appears twice (independently in D and H) and is not synapomorphic. Character 2 appears three times: twice independently in F and I, but also uniting the members of the group ABC. Hence some of its similarity is synapomorphic. (After Goloboff 1991.)

Table 5.2 Example demonstrating the use of the retention index with nine taxa (A–I) and two characters (C1 and C2). Abbreviations for s, m, g, ci and ri as Table 5.1.

			Taxa						s	m	g	ci	ri
A	B	C	D	E	F	G	H	I				(m/s)	$(g - s/g - m)$
C1	0	0	0	1	0	0	0	1	0	2	1	2	0.5	$2-2/2-1=0$
C2	1	1	1	0	0	1	0	0	1	3	1	5	0.3	$5-3/5-1=0.5$

Character 2 has more homoplasy than character 1, which is reflected in their ci values (0.33 and 0.5 respectively). However, character 2 does contribute some synapomorphy to the cladogram and hence contains evidence of grouping, whereas character 1 does not. This is reflected in their respective ri values of 0.5 and 0 (see summary in Table 5.2).

5.1:6 Ensemble retention index (RI)

As for the CI, the ensemble retention index (RI) can be found by using the summed values of m, s and g (M, S and G respectively). For Matrix 1 (Table 5.1a), RI $= G - S/G - M = 10 - 8/10 - 6 = 0.50$, while for Matrix 2 (Table 5.1b), RI $= 13 - 8/13 - 6 = 0.71$. Matrix 2 has an RI value that is a little better (0.71) than that of Matrix 1 (0.50). This is due to more of the similarity being interpreted as synapomorphy. However, as the level of homoplasy is unchanged, so is the CI.

5.2 CHARACTER WEIGHTING

The idea that characters must be or need to be weighted is persistent in cladistics. The first approaches to character weighting were largely subjective. For morphological characters, criteria such as their structural complexity, constancy among taxa, the 'Darwin principle' (characters of low adaptive value should be weighted higher) and character correlation were considered appropriate (Mayr 1969). Pheneticists initially eschewed weighting altogether, while others suggested a variety of approaches not too dissimilar to those outlined by Mayr (1969).

In the cladistic literature, the first serious attempt to incorporate character weights was proposed by Farris (1969). However, they were not explored in detail until the discovery that analysis of a given matrix might result in more than one most parsimonious cladogram. The question was then proposed: given a suite of equally most parsimonious cladograms from one data set, is there a *rational* way to choose one from among the many? Two possible approaches have been suggested:

• As all the equally most parsimonious cladograms are equally 'true', one should not seek to choose from among them but simply represent the common information in a consensus tree.

- The analysis could be repeated incorporating some justified approach to character weighting, in the hope that one cladogram (or at least a smaller subset) would emerge as better supported.

The use of consensus trees for summarizing information from suites of cladograms is discussed fully elsewhere (see Chapter 7). Of significance here is a distinction between how consensus trees can and have been used. Conventionally, the legitimate use of consensus techniques is for summarizing information from a suite of cladograms derived from the same data set, rather than from suites of cladograms derived from the different data sets (although this topic is by no means resolved, and relates to the 'taxonomic' versus 'character' congruence debate; see Chapter 8). However, a consensus tree is generally less resolved than any of the equally most parsimonious cladograms from which it is derived and is therefore considered a poorer summary of the data than any of these original cladograms. For this reason, Carpenter (1988) rejected consensus approaches as the final summary and suggested that prior to consensus tree construction, exploration of character weighting might prove useful.

5.2.1 Types of character weighting

Weighting of characters has been subdivided into *a priori* and *a posteriori* approaches, that is, weighting can be applied before or after cladogram construction. Alternative terms have been proposed: 'hypothesis dependent' and 'hypothesis independent' (Sharkey 1989), and 'tree dependent' and 'tree independent' (Sharkey 1993). However, *a priori* weighting is hypothesis and tree independent, while *a posteriori* weighting is hypothesis and tree dependent. The later terms are therefore redundant and we will use *a priori* and *a posteriori* throughout this discussion.

For *a priori* weighting, there are two different approaches: character analysis and character compatibility. There are also two approaches to *a posteriori* weighting, both based on the notions of cladistic consistency and character reliability: successive approximations weighting (Farris 1969, 1989, hereafter called 'successive weighting'); and implied weighting (Goloboff 1993). While there are other approaches to weighting characters, these have the greatest relevance to cladistic practice.

5.2.2 *A priori* weighting

Character analysis

Character analysis refers to the re-examination of original data in an effort to discover whether any mistakes have been made, such as badly formulated hypotheses of primary homology (similarity) or inappropriate coding. The first type of error, mistaken judgements of similarity, closely resembles Hennig's primary suggestion for resolving incongruence, that of checking and re-checking characters. As a general rule, it must always be considered appropriate to weed out poor delimitation of characters. As such, preliminary cladistic analyses should indicate those characters that perform poorly (defined below) and act as a guide to those that may require re-evaluation. However, even in the most careful studies some character conflict will remain. A good example is DNA sequence data, in which only so much re-sequencing can be done before one must conclude that the conflict in the data is a fact. Therefore, while it should be unreservedly stressed that examination (and re-examination) of specimens (or sequences) is of vital importance, this will probably not eliminate all character conflict.

Below we treat morphological and molecular data separately from the point of view of character analysis. This should not be construed as recognition of different classes of data. On the contrary, there is much that is similar in the analysis of molecular and morphological data. However, molecular data have a fixed number of 'attributes': nucleotide sequences are represented by only four characters (the bases, plus possibly a fifth for 'gaps') and protein sequences by 20 (the amino acids). Hence, regularities might be more easily discovered than in morphological data, as it is possible to calculate accurately the possible permutations of the characters and differential weighting of any empirical 'imbalances' deemed worthwhile.

Morphological data

Neff (1986) presented the most complete analytical method for character analysis to date, broadening the concept to include factors other than re-examination of morphological comparisons. Neff divided the process of phylogeny reconstruction into two distinct parts: character analysis and cladistic analysis (an idea reinvented by Brower and Schawaroch 1996). According to Neff, character analysis involves two steps rather than the one that is generally assumed (character delimitation). Step 1 is synonymous with character delimitation and includes the initial investigation of specimens, making of observations and identification of features as possible homologues. As it pertains to a particular hypothesis, this step is summarized as: 'Feature X in taxon A is the same as feature X in taxon B'. Neff's step 1 is analogous to Patterson's (1982) concept of similarity and Rieppel's (1988) idea of 'topographic correspondence' being the initial criterion for homology determination. Neff's second step consists of constructing a hierarchy of characters and is also phrased with reference to particular hypotheses: 'Feature X is more general than, and includes, the more specialized feature Z'. Step 2 thus concerns polarity estimation (this subject was considered in detail in Chapter 3). For the purposes of this discussion, polarity decisions should be understood to be conclusions that are derived from a cladogram and that cladogram construction does not require *a priori* polarization (see Chapter 3). Yet there is much of merit in Neff's ideas. Indeed, as Carpenter (1988: 292) noted: 'Actually Neff's paper [1986] merely argues for careful homology decisions, which should be given'.

Molecular data

Judgemental mistakes of similarity are a more than reasonable source of error when considering morphological data. This is as it should be. Examination of character conflict leads to re-examination of characters, which, in turn, leads to greater understanding of the organisms and improved classifications. This is the essence of systematics. Yet when dealing with sequence data (particularly nucleotide sequences), one is usually faced with the conclusion that the data do, indeed, contain copious 'real' conflict. This is because it is impossible to examine a nucleotide base in any further detail, its similarity is as exact as can be. However, the regularity that defines sequence data may be used to implement certain types of *a priori* weighting, usually on the assumption that the conflict is 'caused' by known processes (but see below). The various types of weighting possible are listed as follows (Hillis *et al.* 1993); perhaps more will yet be discovered.

A priori weighting:

  Uniform weighting (all bases given equal weight)

  Non-uniform weighting (bases assigned different weights)

    Across positions (structural-functional differences)

      Codon positions (selective weighting of first, second and third positions in relation to redundancy of genetic code)

      Stems and loops (selective weighting of loops versus stems in secondary folding structure (RNAs))

    Within position (mutational bias)

      Transversions versus transitions (weighting of transitional bias)

      Relative substitution base composition (12 possible substitutions weighted according to observed or expected frequencies)

      Synonymous versus non-synonymous change (change of amino acids in coding regions)

A posteriori weighting:

  Successive approximations weighting (weights according to levels of homoplasy)

  Dynamic weighting (weights according to levels of homoplasy, includes within and across positions)

Weighting nucleotide sites can be complex, depending on what is considered to be significant in the data. The 'default' approach is uniform weighting, in which all bases are given equal weight. Non-uniform weighting selects some differential that can be assessed from the data prior to analysis. This can involve across sequence and within sequence positions or combinations of both.

Across sequence position weighting involves known structural or functional differences in a particular molecular sequence. The best known example relates to the degeneracy of the code. The synthesis of proteins requires the genetic information to be translated into the correct amino acids, mediated by transfer RNA (tRNA). Each tRNA recognizes a particular triplet of nucleotides (codon) which represents a particular amino acid. As there are four different nucleotides and each codon is a triplet, there are 64 possible codons. Of these, 61 code for amino acids, while the remaining three code for 'nonsense' or 'stop' codons, which bring translation to a halt. With only 20 amino acids, some of the 61 codon triplets must code for identical amino acids. In this sense the code is degenerate. Different codons that represent the same amino acid are called synonymous. Therefore, a common method to weight coding sequence data is to weight differentially the first, second and third positions of the codon relative to particular amino acids. For example, the amino acid proline is coded by four codons: CCU, CCC, CCA, CCG. Any substitution in the third position will not result in a change of amino acid and all third position changes for proline are synonymous. Likewise, leucine is coded by six codons: UUA, UUG, CUU, CUC, CUA and CUG. If a codon has the form CUX (where X stands for any base), the amino acid will always be leucine, and all third position changes are synonymous. However, the codons UUA and UUG also code for leucine. If the codon UUX has a third position substitution, then codons UUU and UUC will result in phenylalanine. Thus, for leucine and phenylalanine, not all third position changes are synonymous. Some are non-synonymous in that some substitutions result in different amino acids. It is possible to calculate all the different kinds of changes that can occur in coding sequences. For instance, as there are 61 sense codons there are 549 possible nucleotide substitutions. Thus, it is possible to calculate all of the different kinds of possible amino acid coding and the relative frequency of the change for each codon position. Such calculations show that 70% of third position changes are synonymous, all substitutions at the second position are non-synonymous, and 96% of substitutions at the first position are non-synonymous. It thus appears rational to downweight or even ignore third position changes. The most extreme form of this type of weighting is to use the amino acid sequence as the primary data and not the nucleotide sequence (e.g., some *rbc*L studies). While statistically the frequency of changes expected at individual codon positions is clear, empirical examination of the effects of differential weighting may reveal different kinds of relationships (e.g. Allard and Carpenter 1996). It seems

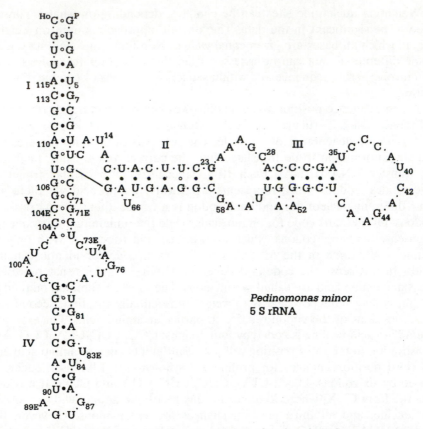

Fig. 5.4 Secondary structure in the 5S rRNA molecule of *Pedinomonas minor*. Paired sites are those which form Watson–Crick base pairs in the stem regions of a molecule and unpaired sites are those that occur in the loop regions. (After Devereux *et al.* 1990.)

wise to examine each molecule individually rather than attempt to extrapolate a particular situation into a generality pertaining to all molecules.

With respect to non-coding genes, different factors can be taken into account. For instance, every molecular sequence has both a secondary and tertiary folding structure such that some bases are placed adjacent to each other in stem regions and others are separate in loop regions (Fig. 5.4). Nucleotide bases that appear opposite each other in stem regions are seen as dependent, because if a substitution occurs in one position, the opposite base may also have to change to maintain the overall structure. In contrast, bases in the loop regions have fewer such constraints and may be free to change to any other nucleotide. Therefore, it might seem useful to weight these positions accordingly. However, disagreements exist with respect to the informativeness of the stem and loop regions. For instance, Wheeler and Honeycutt

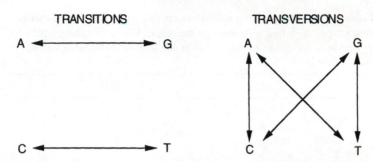

Fig. 5.5 Classification of nucleotide substitutions. Each arrow represents two options for direction of change. Transitions, of which there are four possibilities, substitute one purine (A or G) for another or one pyrimidine (C or T) for another. Transversions, of which there are eight possibilities, substitute a purine for a pyrimidine or vice versa.

(1988) suggested that the stem regions were uninformative (and effectively gave them zero weight), while Dixon and Hillis (1993) came to the opposite conclusion. They suggested that the informativeness of the stem and loop regions may be unique to each molecule or to particular organisms (Wheeler and Honeycutt examined 5S and 5.8S rRNA genes in insects, while Dixon and Hillis used 28S rRNA in vertebrates). Such discordant conclusions suggest that even if it appears legitimate to implement this type of weighting scheme, it may not be one that can be generalized to all organisms and all molecules. Once again, the 'imbalances' need to be investigated with reference to each particular problem and their utility examined against the conclusions (a cladogram).

Within-sequence position weighting utilizes several different kinds of proposed mutational bias. The first to be explored was the relative frequency of transitions versus transversions. There are two kinds of nucleotides: adenine (A) and guanine (G) are purines, while cytosine (C) and thymine (T) are pyrimidines. Transitions substitute one purine for another or one pyrimidine for another, of which there are four possibilities (Fig. 5.5). Transversions are substitutions of a pyrimidine by a purine or vice versa, of which there are eight possibilities (Fig. 5.5). By chance alone, one would expect to encounter more transversions than transitions. However, in many examples, there are significantly more transitions than transversions. Such imbalances can be used to weight the data so as to prevent favouring one kind of substitution over another. For example, Miyamoto and Boyle (1989) discovered that 'better results in terms of unambiguous resolution..., congruence..., and consistency... are expected from analyses of transversions alone, rather than from combinations of transitions, transversions, and gap events...'

Of further significance are the relative substitution frequencies of all 12 possible substitutions, which can be weighted according to their frequency in any particular set of sequences. There are 16 possible substitutions, of which

Table 5.3 The 16 possible substitutions among the four bases with frequencies *a–l*; frequencies *w–z* represent the four 'substitutions' of one base by the same base, which are thus undetectable and are disregarded.

	A	C	G	T
A	*w*	*a*	*b*	*c*
C	*d*	*x*	*e*	*f*
G	*g*	*h*	*y*	*i*
T	*j*	*k*	*l*	*z*

four are substitutions of one base with one that is identical and hence are undetectable and disregarded (Table 5.3, $w - z$). Of the 12 observable changes, different frequency values are allowed for each direction of change, such that $A \rightarrow C$ is *d* and $C \rightarrow A$ is *a*, where *d* and *a* may be different. These values can be calculated from observed frequencies and compared against expected frequencies. The values can be used in a variety of ways, including step-matrices (see Chapter 4).

It is possible to combine within and across sequence weighting regimes, as in dynamic weighting (Williams and Fitch 1990, Fitch and Ye 1990), which is based, in part, on 'successive weighting' (see below). In successive weighting, weights are assigned according to their levels of homoplasy on resulting cladograms. Dynamic weighting applies the same strategy but also includes information on the relative frequency of the observed character change. As an example, Marshall (1992) used dynamic weighting to re-analyse small subunit (SSU) rRNA sequences for amniotes. In the original analysis, Hedges *et al.* (1990) found support for a sister-group relationship between birds and mammals to the exclusion of crocodiles (Fig. 5.6b; a solution favoured by some morphologists, e.g. Gardiner 1993). However, inspection of the data revealed considerable substitution bias. In particular, there was an over-representation of T–C substitutions and a significant under-representation of A–T and T–A substitutions. Marshall noted that a large number of sites with T–C substitutions supported the bird–mammal relationship. Using dynamic weighting, Marshall discovered instead that the data supported a bird–crocodile sister-relationship (Fig. 5.6c, differing slightly from the traditional 'palaeontological' cladogram, Fig. 5.6a). These results may only attain significance with respect to a wider analysis of all molecular and morphological (including palaeontological) data (as in the comprehensive study by Eernisse and Kluge 1993, which suggested little overall support for the bird–mammal relationship).

It is often taken as fact that it is more reasonable to apply *a priori* weighting schemes to sequence data. Many, if not all, such weighting schemes are conceived in terms of kinds of substitution. However, the notion of substitution (a process) is itself derived from the data. Comparison of two (or

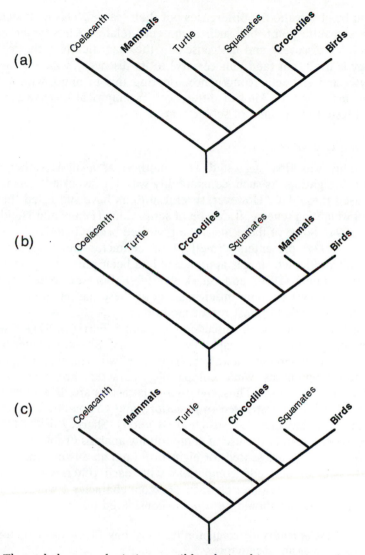

Fig. 5.6 Three cladograms depicting possible relationships among amniotes. (a) The palaeontological cladogram. (b) The cladogram derived from the unweighted 18S rRNA data of Hedges *et al.* (1990). (c) The cladogram derived from the weighted 18S rRNA data of Marshall (1992).

more) sequences will either reveal identity or not at particular sites. The difference is empirical (derived from a comparison). Calling the differences between two sites 'substitutions' may simply be a label based upon presumed understanding of the 'cause' of the difference. Yet the actual difference and its cause can be seen as two different things. In this sense, weighting is arrived at by observation of differences not their 'cause'. This may seem to be a minor semantic point, but such arguments relate to the larger issue of homology, its discovery and its 'cause'. In the cladistic view, the 'cause' of homology is irrelevant (and unnecessary) to its discovery, which is the result of analysis and character congruence. Bearing this in mind, with sequence data one may view the kind of differences as empirical observations to be seen and tested in the light of subsequent analysis.

Compatibility analysis

Compatibility was first suggested as a method of analysis rather than a method of weighting. As such, compatibility was not favourably received and is little used these days. However, several authors have suggested that when used as a weighting scheme, it may be of some value (Penny and Hendy 1986, Sharkey 1994). Some of these ideas are reviewed briefly below.

Sharkey (1994) presented a method of character weighting that used compatibility, although similar approaches had been presented previously by Penny and Hendy (1986) and Sharkey (1989) (but see Wilkinson 1994). However, Sharkey had some misgivings about these earlier attempts, hence we will consider only his most recent method.

Sharkey (1994) began his discussion by citing Farris' (1971) distinction between 'congruent' and 'compatible' characters. Sharkey's interpretation suggested that congruent characters correlate with respect to particular phylogenetic hypotheses, while compatible characters are correlated with each other in the data set. Thus, congruent characters are determined after a cladogram has been constructed (*a posteriori*) and compatible characters are judged prior to cladogram construction (*a priori*). Sharkey devised a simple example to demonstrate the use of compatibility analysis (Table 5.4), in which 13 binary characters are scored for eight taxa (and an all-zero root). Characters 1–12 are all perfectly compatible with each other. Character 13 is incompatible with every other character (except character 1 which is uninformative) and therefore should be *a priori* considered the weakest in the data

Table 5.4 Character matrix for 'compatibility' weighting. Characters 1–13 (columns), taxa A–H with O as an all-zero root.

	1	2	3	4	5	6	7	8	9	10	11	12	13
O	0	0	0	0	0	0	0	0	0	0	0	0	0
A	1	1	1	1	1	1	1	1	1	1	1	1	0
B	1	1	1	1	1	1	1	1	1	1	1	1	1
C	1	1	1	1	1	1	1	1	1	1	1	1	0
D	1	1	1	1	1	1	1	1	1	1	1	1	1
E	1	1	1	1	1	0	0	0	0	0	0	0	0
F	1	1	1	0	0	0	0	0	0	0	0	0	0
G	1	1	0	0	0	0	0	0	0	0	0	0	1
H	1	0	0	0	0	0	0	0	0	0	0	0	1

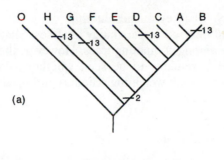

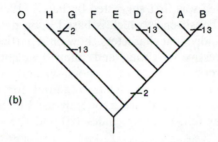

Fig. 5.7 The two equally most parsimonious cladograms derived from the data in Table 5.4. (a) On this cladogram, character 2 fits perfectly with a single step, while character 13 fits with four steps. (b) In contrast, on this cladogram, character 13 fits with only three steps, although character 2 now fits less than perfectly with two steps.

set. Parsimony analysis yields two cladograms (Fig. 5.7). The first cladogram (Fig. 5.7a) accounts for character 13 with four occurrences (branches leading to taxa B, D, G, and H). This is its worst possible performance on any cladogram. The second cladogram (Fig. 5.7b) accounts for character 13 with three occurrences (branches leading to taxon G + H, and branches leading to B and D). Character 13 fits the second cladogram better than the first by one step and provides additional support for the group G + H from character 3 (as a 'reversal'). Both cladograms are equally parsimonious. In this case, the first cladogram (Fig. 5.7a) is preferred because more characters are compatible. As Sharkey pointed out, this is not based upon any *a posteriori* consideration of fit, but on the *a priori* consideration of compatible characters 1–12. It is worth noting that the first cladogram would also be selected by successive weighting (see below).

To measure the relative amount of compatibility in any data set, Sharkey proposed the unit discriminate compatibility measure (UDCM) of a character, described as 'the complement of the probability of a derived character state being nested with another derived character state or the probability of a derived character state being exclusive of another derived character state, depending on the observed pairwise character comparison' (Sharkey 1994). The UDCM allows weights to be assigned to characters on the basis of their overall compatibility. Sharkey suggested that once the weights had been calculated they could be used in parsimony analyses. Sharkey's objective was to assist in choosing among competing trees. However, the idea has not been tested in detail and it remains to be seen whether, like compatibility for cladogram construction, such weighting schemes end up discarding (by down-weighting) many characters.

5.2.3 *A posteriori* weighting

A posteriori weighting refers to methods that derive weights after cladogram construction. The idea was first suggested by Farris (1969) in relation to the concept of cladistic consistency, which is based upon measuring the amount of discordance that individual characters show on a particular cladogram, and is the basis for 'successive weighting' and 'implied weighting'. Discordance is another way of describing levels of homoplasy, and homoplasy is usually reflected by the number of extra steps required for characters to fit a cladogram. Measures of discordance frequently used in cladistic analyses are the consistency index (ci), retention index (ri) and the rescaled consistency index (rc).

Cladistic consistency

The essence of cladistic consistency can be grasped from a simple example taken from Farris (1989). Consider the fit of two characters, 1 and 2, on two equally parsimonious cladograms, X and Y. Character 1 is a binary character which, if it fits any cladogram in the worst possible way, will have only two steps. Character 2 is also a binary character which, if it fits any cladogram in the worst possible way, will have 15 steps. Character 1 fits cladogram X with 1 step (its best) and fits cladogram Y with 2 steps (its worst). Character 2 fits cladogram X with 15 steps (its worst) and fits cladogram Y with 14 steps (nearly its worst). Both cladograms are equally parsimonious and only one step is saved in each character. One may reasonably ask, which of the two cladograms, if either, is to be preferred? It would seem intuitively obvious that cladogram X is the better of two. The fit of character 1 is perfect (1 step, character 1 is a synapomorphy) and while character 2 has a worst-case fit of 15 steps, the improvement of a single step achieved on cladogram 2 is not very impressive. Character 2 is clearly less informative than character 1 with reference to both cladograms X and Y. Therefore, while one step is saved on each cladogram, it seems preferable to use those characters which have an overall better fit. In other words, characters should get the weight they deserve. Such ideas suggest that some characters are cladistically more reliable than others. Successive weighting incorporates this idea, although its implementation was not widely available until the release of the parsimony program, Hennig86. The implementation of successive weighting in Hennig86 differs from that originally outlined by Farris (1969) and is described below.

Successive weighting

First, we need to recapitulate briefly the measures of homoplasy. When characters fit a cladogram perfectly (that is, for binary characters, to a single node only) they are 100% consistent with that cladogram. When characters do not fit a cladogram perfectly (for binary characters, to more than one node) they are less than 100% consistent. Characters that do not fit to one node of a cladogram exhibit homoplasy, which is discordance with a particular cladogram and is measured by the consistency index, ci. Thus, the cladistic consistency of a character is potentially measurable relative to its performance on a particular cladogram. However, although a character may show homoplasy on a particular cladogram, this does not mean that all of its similarity need be uninformative. The amount of similarity interpreted as synapomorphy is measured by the retention index, ri. The weight of a character can be seen as a function of its fit to a cladogram and requires consideration of both homoplasy and synapomorphy. Both ci and ri can be used to estimate weights for characters that have a direct bearing on their evidential value with respect to the recovered cladograms. Farris (1989) suggested rescaling ci using the ri values, to give the rescaled consistency index (rc): essentially this is the product of ci and ri. Thus, characters with no similarity interpreted as synapomorphy (ri = 0) will be disregarded, irrespective of their level of homoplasy. Those that have some similarity interpreted as synapomorphy relative to a cladogram will have a weight proportional to the amount of homoplasy.

Referring back to Matrix 1 (Table 5.1a) and the cladogram in Fig. 5.1c, four characters fit this cladogram perfectly with ci = 1 and ri = 1 (characters 1–4, Table 5.5). Their 'deserved' weights are calculated using rc, giving each a value of 1 (ci × ri), which is then scaled by Hennig86 between 0 and 10 to give final weights of 10. (PAUP scales the weights between 0 and 1000.) Characters 5 and 6 fit the cladogram with two steps and thus ci = 0.5. However, none of their similarity is interpreted as synapomorphy and so for both characters ri = 0. Consequently, rc = 0 (ci × ri). They can thus play no part in determining the topology of the cladogram. As the weights can affect both the number and topology of the resulting cladograms, it is necessary to ensure that the weights stabilize. This is achieved by repeating the reweighting procedure until the weights assigned to each character in two successive iterations are identical—hence the name, successive weighting.

Other parsimony programs, such as PAUP, can implement other indices or combinations of indices to assign weights, but the use of such measures is still subject to some debate. However, the principles of successive weighting remain the same.

One major misconception concerning successive weighting must be laid firmly to rest. It is widely believed that successive weighting can be used to reduce the number of equally most parsimonious cladograms found for a given data set and even that this is the prime function of the procedure. This

Table 5.5 Matrix for successive and implied weighting. Characters 1–6, taxa A–D (extracted from Matrix 1, Table 5.1a). l = length, m = minimum possible steps, g = minimum steps on a bush, ci = consistency index, ri = retention index, sw = successive weighting using rescaled ci, iw = implied weighting using $(K/(K + ESi))$, where K = constant and ESi = extra steps. In this example, $K = 3$.

Taxa	Characters					
	1	2	3	4	5	6
A	0	0	0	1	1	1
B	0	0	1	1	1	1
C	0	1	1	1	0	1
D	1	1	1	1	0	0
l	1	1	1	1	2	2
m	1	1	1	1	1	1
g	1	2	2	1	2	2
ci	1	1	1	1	0.5	0.5
ri	1	1	1	1	0	0
sw (ri × ci)	10	10	10	10	0	0
iw ($K/(K + ESi)$)	0	10	10	0	7.5	7.5

it most certainly is not. Successive weighting is a method for selecting *characters* according to their consistency on a given set of cladograms. It is not a method for choosing among those cladograms. If the application of successive weighting results in a smaller number of equally most parsimonious cladograms, then that may be considered a fortunate side-effect. But this need not necessarily happen. If many characters have low consistencies, and thus receive low weights, then the number of equally most parsimonious solutions may actually increase, sometimes quite dramatically. Even if the number of minimum length cladograms does decrease, they still need not be a subset of the original set. Successive weighting effectively creates a new data set, in which some characters (the more consistent) are replicated more times than others (the less consistent). We should not therefore be surprised if successive weighting produces more and different topologies from those with which we started. It also follows that if we are to be consistent in applying successive weighting, we should do so even if we initially obtain only a single most parsimonious cladogram. Furthermore, we should not consider successive weighting to have 'failed' if this one cladogram is then replaced with over 1000 equally most parsimonious solutions, because this result actually implies that the data supporting the original solution were not very consistent and the hypothesis of relationships was not at all strongly supported.

Implied weighting

With respect to homoplasy, extra steps and fitting characters to cladograms,

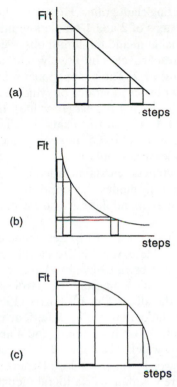

Fig. 5.8 Graphs depicting the three kinds of fitting function used to adjust for relative cladistic consistency. (a) Linear. (b) Concave. (c) Convex. (After Goloboff 1993.)

Farris (1969) discussed three forms of fitting function that could be used to adjust for relative cladistic consistency: linear, concave, and convex (Fig. 5.8). For linear fit (Fig. 5.8a), the cladogram with the overall shortest length is considered optimal. The problem with this approach is that the relative values of the characters are ignored. Linear fitting is used when equal (uniform) weighting is applied, that is, all steps count equally. It is therefore equivalent to the 'default' option of most parsimony programs. One conclusion that can be drawn from uniform weighting is that the reliability of the weights (and hence the characters) is set prior to the analysis—they are all equal. Yet it is clear from many, if not most, analyses that there will always be some characters that behave well and others that behave poorly. The implication is that they do not all contribute equal kinds of information.

For concave fit (Fig. 5.8b), a non-linear relationship reflects how well each character performs on a relative scale. To return to the example given above, which considered two binary characters (1 and 2) and two different but equally parsimonious cladograms (X and Y), both characters differed by a single step on the competing cladograms. However, character 1 has maximum and minimum observed steps of 2 and 1 on cladograms Y and X respectively, while character 2 has maximum and minimum observed steps of 15 and 14 on cladograms X and Y respectively. Intuitively, character 1 should receive greater weight than character 2. The proportional difference can be assessed by use of 'extra steps'. For two cladograms on which a character has s^1 and s^2 steps respectively (with s^1 having the larger value), this proportional difference is given by $(s^1 - s^2)/(s^1 s^2)$. In this example, character 1 has a value of 0.5 and character 2 has a value of 0.005. In short, concave fit gives preference to those characters with least homoplasy.

For the sake of completeness, Farris included a short discussion on convex fit (Fig. 5.8c). This approach implies the opposite of concave fit, suggesting that characters with greater homoplasy are to be preferred—clearly, this is not a sensible option!

Goloboff (1993) exploited concave fit to determine character weights in his computer program, PIWE. Here, weights are calculated as $W = K/(K + ESi)$, where ESi is the number of extra steps per character and K is the constant of concavity (the inclusion of a measure of 'extra steps' makes Goloboff's approach analogous to the discussion of Farris (1969)). Referring back to Matrix 1 (Fig. 5.1a) and cladogram in Fig. 5.1c, four characters (1–4) fit this cladogram perfectly (Table 5.5). Characters 1 and 4 are uninformative and so PIWE assigns them zero weight, in contrast to successive weighting which assigns them maximum weight (in practice, of course, this makes no difference to the analysis, merely adding to cladogram length). Characters 2 and 3 fit the cladogram perfectly, that is with no extra steps, and hence receive the maximum weight of 10. Characters 5 and 6 fit the cladogram with two steps and hence each character has one extra step. Thus, for $K = 3$, they receive weights of 7.5. Two consequences can be seen immediately from this simple example. All characters, unless completely uninformative, receive a non-zero weight and this weight varies according to the value assigned to K. When K is altered, the weights change. For characters 5 and 6, weights for values of $K = 1$ to 6 are given in Table 5.6. Goloboff (1993) then used the total weight of all characters, rather than cladogram length, to select the best cladogram.

Table 5.6 Weights for characters 5 and 6 from Table 5.5 with $K = 1$–6.

Value of K	Weight assigned
1	$1/(1 + 1) = 0.50 \times 10 = \mathbf{5.0}$
2	$2/(2 + 1) = 0.66 \times 10 = \mathbf{6.6}$
3	$3/(3 + 1) = 0.75 \times 10 = \mathbf{7.5}$
4	$4/(4 + 1) = 0.80 \times 10 = \mathbf{8.0}$
5	$5/(5 + 1) = 0.83 \times 10 = \mathbf{8.3}$
6	$6/(6 + 1) = 0.85 \times 10 = \mathbf{8.5}$

In Goloboff's terminology, the 'heaviest' cladogram is selected by summing all the assigned weights and choosing the cladogram with the largest value.

Implied weighting is relatively new and few studies have yet been undertaken (Goloboff 1995b, Szumik 1996 and, for critical examination, see Turner 1995).

5.2.4 Prospects

The mechanics and implementation of *a posteriori* weighting are described above, but what of the results? Carpenter's general aim was to investigate a rational way to reduce the number of equally parsimonious cladograms using empirical criteria. However, results suggest that this general expectation need not be the case, for either successive weighting or implied weighting. Two significant results have emerged. First, a greater rather than a smaller number of cladograms than were in the original set may be recovered after weighting, and second, the cladograms found after weighting may not be among the original equally most parsimonious suite (e.g. Platnick *et al.* 1991*b*).

With respect to more cladograms, it may be that of the original equally weighted characters, few were cladistically consistent (or self-consistent in the terminology of Goloboff 1993) and many received zero weight. The cladograms derived from the equally weighted data thus depend upon a few ambiguous characters. The significance of obtaining different topologies from *a posteriori* weighted data is perhaps more controversial. It has been argued that this result is not unexpected (Platnick *et al.* 1991*b*, Goloboff 1993), for, in practical terms, weighting is equivalent to excluding some characters and non-randomly replicating others. Consequently, a differentially weighted data set need not give the same results as an equally weighted one. Others have argued quite simply that longer cladograms should be disregarded as they violate the basic premise of parsimony (Turner and Zandee 1995).

Consider one example. The elegant study of haplogyne spiders by Platnick *et al.* (1991*a*) yielded ten equally most parsimonious cladograms using equal weights (length = 184). After successive weighting (implemented by Hennig86), this number was reduced to six (length 568; note this large increase in length is to be expected, because, for example, a perfectly consistent character now adds 10 to the length of the cladogram, rather than the previous 1). None of the six weighted cladograms were among the original ten. When the six weighted cladograms were inspected using the unweighted matrix, two cladograms had a length of 185 (one step longer than the optimal cladograms from equal weighting) and the other four had a length of 187. Platnick *et al.* reasoned that of the six cladograms, the two with length 185 were worthy of further consideration and were better than the equally weighted cladograms in spite of the fact that they are one step longer. They argued that for the weighted cladograms, the characters contributing to the topology could be considered more 'consistent' than those under equal weights. This suggests a further conclusion relevant to weighting. Rather than being a method to *reduce* the number of equally parsimonious cladograms, Goloboff (1993) and Platnick *et al.* (1991*a*) have suggested that parsimony analyses *require* weighting to achieve self-consistent results, even if only a single most parsimonious cladogram is found using equal weights (this view has been contested by Turner and Zandee 1995; with a reply by Goloboff 1995*a*). Platnick *et al.* (1996) have further suggested that equal weights can only be considered a preliminary and crude estimate of the relative value of the data. Such views are consistent with the general understanding of cladistic parsimony, that the 'value' of a character is related to its performance on a cladogram. Further, coupling this with a notion of support for each clade leads to a firmer choice of particular cladograms, whether only one results from analysis or many.

A posteriori character weighting holds a certain amount of promise and may help to produce more consistent cladograms relative to the data collected. Progress may result from performing analyses with different parameters for weighting (as well as different values for obtaining weights) on more data sets (e.g. Suter 1994). Whatever the outcome, it seems likely that the view of Platnick *et al.* (1996), that equal weights can only be a preliminary and crude estimate of any particular data set, is worthy of further consideration.

5.3 CHAPTER SUMMARY

1. The simplest measure for assessing the fit of data to a cladogram is cladogram length. Being the shortest, the most parsimonious cladogram has the best fit to the data. Longer cladograms have poorer fit.

2. The consistency index, ci, of a character is defined as the ratio of m, the minimum number of steps a character can exhibit on any cladogram, to s, the minimum number of steps the same character can exhibit on the cladogram in question. The consistency index measures the amount of homoplasy in the data. Problems with the consistency index are that unique characters (autapomorphies) and invariant characters will inflate its value; it can never attain a zero value; and, in general, its value is inversely proportional to the number of taxa included in the analysis.

3. The retention index, ri, was introduced to address the problem of character fit to a cladogram, as opposed to amount of homoplasy displayed by a character. The ri is defined as $(g - s)/(g - m)$, where g is the greatest number of steps a character can exhibit on any cladogram. The retention index measures the amount of similarity interpreted as synapomorphy and, unlike the ci, can attain a value of zero.

4. Values for ci and ri are useful for examining how individual characters perform on a cladogram. In order to assess how an entire data set performs, the ensemble consistency index (CI) and ensemble retention index (RI) are used. $CI = M/S$ and $RI = (G - S)/(G/M)$, where M, S and G are the sums of all the m, s and g values for the individual characters respectively.

5. Weighting of characters can be either *a priori* (tree independent, hypothesis independent) or *a posteriori* (tree dependent, hypothesis dependent).

6. There are two types of *a priori* weighting. A character can be weighted according to what we can discover about its origin, either empirically (ontogeny) or historically (phylogeny). Phylogeny differs from ontogeny in that it requires knowledge of the 'evolutionary' properties of characters, which may not be easily or even readily accessible. This kind of weighting is usually understood as some kind of connection between parsimony and evolution. 'Character analysis' is considered by some to be a form of weighting but simply refers to a thorough re-examination of data from specimens to ensure that structures are worthy of comparison. Alternatively, characters can be understood as being 'compatible' with each other and may suggest that particular associations of taxa can be *a priori* prohibited. Character compatibility has had a somewhat chequered history, having been used as a method of phylogenetic reconstruction and consensus tree construction, as well as *a priori* character weighting.

7. Weights can be assigned *a posteriori* using the concept of character consistency, a concept directly related to the cladograms resulting from a cladistic analysis and their implied levels of homoplasy.

8. The consistency index can be used to assign weights to characters. However, because the ci cannot attain a zero value, Farris suggested rescaling the ci using the retention index, to produce the rescaled consistency index, rc.

9. Once weights have been assigned to characters using rc, the most parsimonious cladograms for the weighted data set are found. The cladistic consistency of the characters on these new cladograms forms the basis of a new set of weights, and the process is continued until the weights stabilize. This is successive approximations character weighting.

10. In contrast, Goloboff suggested that the weight of a character (its 'implied weight') is a function of its fit to a cladogram. Therefore, the best cladogram maximizes fit, giving a measure of 'total fit'. The function of fit to a cladogram requires consideration of fit in terms of homoplasy.

11. Farris suggested three kinds of fitting function: linear, concave and convex. Goloboff recommended concave fit (following Farris) and adopted an approach that considers the direct weighting of 'extra steps', such that the weight of a character is given by $K/(K + ESi)$, where ESi is the number of extra steps and K is the constant of concavity.

6.
Support and confidence statistics for cladograms and groups

6.1 INTRODUCTION

The study of phylogeny is an historical science, concerned with the discovery of historical singularities. Consequently, we do not consider phylogenetic inference *per se* to be fundamentally a statistical question, open to discoverable and objectively definable confidence limits. Hence, we are in diametric opposition to those who would include such a standard statistical framework as part of cladistic theory and practice. However, while the most parsimonious cladogram does represent the best summary of the data to hand and is thus the preferred hypothesis of relationships among the study taxa, it is naïve to assume that it also represents the 'true phylogeny'. Cladograms are always subject to revision in the light of new data, reinterpreted hypotheses of homology or improved analytical methods. But such changes can lead to instability in taxonomy and nomenclature if they are undertaken too frequently or without sufficient forethought. If authoritarianism is to be avoided, some objective means needs to be found to determine which changes are to be included in the new, improved classification and which are not.

6.2 RANDOMIZATION PROCEDURES APPLIED TO THE WHOLE CLADOGRAM

A question often asked of a data set is whether it contains 'significant cladistic structure', that is, whether we can have any confidence that the results of a cladistic analysis are, in some sense, 'real' and not just by-products of chance. The concept of cladistic structure can be studied from two viewpoints. The first attempts to assign confidence to the most parsimonious cladogram as a whole, while the second examines the support afforded to individual clades within the most parsimonious cladogram and asks which of these are reliably supported by evidence and which only weakly so.

To attempt to answer the question 'Could a cladogram as short as this have arisen purely by chance?', it is necessary to compare the length of the most parsimonious cladograms derived from the real data with those obtained from 'phylogenetically uninformative' data sets. Several definitions of phylogenetically uninformative data have been proposed. For example, Archie and

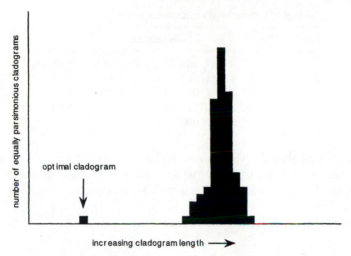

Fig. 6.1 The basic concept underlying most tree support statistics. A data set of real observations is repeatedly perturbed according to a particular set of rules to yield a large number of pseudoreplicate sets of 'phylogenetically uninformative' data. The length of the most parsimonious cladogram(s) derived from the real data set is then compared with the lengths of the most parsimonious cladograms obtained from these contrived data sets, with the expectation that the former (indicated by the arrow) would be very much shorter than any of the latter (represented by the frequency histogram).

Felsenstein (1993) interpreted it to mean 'statistically random'. In this method, random data are generated using a model in which there is an equal probability of state 0 or 1 in every cell. A random addition sequence algorithm is then used to generate 'random cladograms'. The expected number of steps on each random cladogram is then calculated. These values are then compared with the length of the most parsimonious cladogram obtained from the real data set, with the expectation that the latter should be substantially shorter (Fig. 6.1).

6.2.1 Data decisiveness

In contrast, Goloboff (1991) argued that systematists expect data sets to contain information that allows them to choose among different hypotheses of relationship. A data set in which all possible informative characters occur in equal numbers will provide no such evidence because all possible fully resolved topologies will be of the same length. Because no choice can be made among the cladograms, the data set is phylogenetically uninformative. Goloboff termed such a data set 'undecisive' (Table 6.1). In contrast, a decisive matrix yields cladograms that differ in length among themselves and thus offers a reason for choosing some cladograms over others. As the degree of difference in length increases, so does the decisiveness of the matrix.

Table 6.1 Example of an undecisive data set

Taxa	Characters
A	110 110 1 100 110 110 1 100 1 1000
B	101 101 1 010 101 101 1 010 1 0100
C	011 011 1 001 011 011 1 001 1 0010
D	000 111 0 111 000 111 0 111 1 0001
E	000 000 0 000 111 111 1 111 0 1111

Goloboff showed that the CI, RI and RC do not vary directly with decisiveness and hence the least decisive matrices are not necessarily those with the lowest values for these statistics. Nevertheless, a general measure of decisiveness is possible.

Data decisiveness (DD) is defined as:

$$DD = \frac{\bar{S} - S}{\bar{S} - M}$$

where S is the observed length of the most parsimonious cladogram, $\bar{S}$ is the mean length of all possible bifurcating cladograms and M is minimum possible length of a cladogram were there no homoplasy in the data (the same variable, M, that is used in the calculation of RI). The higher the value of DD, the more the cladograms derived from a data matrix differ in length. DD decreases as characters increasingly conflict with one another, with a value of 1 implying there is no conflict, while a value of 0 is achieved for wholly undecisive data. It is unaffected by the presence of uninformative characters in the data. DD is not necessarily correlated with the number of equally most parsimonious cladograms for a data set, nor has it any strict connection with the strength of preference for the most parsimonious cladogram over every other cladogram. A very low decisiveness implies that there are only very weak reasons to prefer the most parsimonious cladogram over any other topology, including those just one step longer. However, the converse does not necessary hold true, that a high DD signifies a high cladistic information content. In particular, DD cannot be used to assign a confidence level to the most parsimonious cladogram. To achieve this, it would be necessary to compare whether the observed decisiveness is significantly lower than that of random data with the same number of characters and thus far, the statistical distribution of decisiveness generated from random data remains undetermined.

6.2.2 Distribution of cladogram lengths (DCL)

Several authors have examined the distribution of cladogram lengths (DCL; strictly, the distribution of lengths of all possible bifurcating cladograms) for a given data set as an indicator of phylogenetic signal in the data. They have argued that for a DCL that is nearly symmetrical, many cladograms will be only a few steps longer than the most parsimonious cladogram and the phylogenetic signal is weak. However, if the DCL is strongly negatively or left-skewed (i.e. with a long tail to the left side of the distribution), then there are relatively few cladograms that are just slightly longer than the most parsimonious solution and the phylogenetic signal is therefore strong.

One significance test proposed by Hillis (1991) is based upon the null model in which characters are generated independently and at random, and all states have the same expected frequency. We would then conclude there is significant cladistic structure, and that the characters are highly congruent, when the skewness statistic, g_1, for the real DCL is below the fifth percentile (for example) of the DCL g_1 derived from matrices of randomly generated data. However, this effect can also result from the characters simply having different frequencies among the taxa. A character that divides the taxa into two groups of similar size tends to make the DCL more symmetrical. Conversely, characters that allow the recognition of small groups of taxa tend to increase left-skewness. The significance test outlined above confounds this effect with that due to character congruence.

Huelsenbeck (1991a) compared the DCLs generated from real and random data using g_1. Using simulation tests, he found that data that were consistent with only a single most parsimonious cladogram tended to produce a strongly left-skewed DCL. In contrast, data that were consistent with numerous equally most parsimonious cladograms produced a more symmetrical DCL that could not be distinguished from those generated from random data. However, a hierarchical pattern will be recovered whenever there is character congruence, whatever its source, and g_1 is generally negative even for randomly generated data because of chance congruence within such data. Thus, the test is whether the observed skewness in DCL is no less negative than would be expected from random data. If it is, then the conclusion is that there is significant phylogenetic (hierarchical) signal in the data. If it is not, then one would conclude that the observed character congruence is largely due to chance and the most parsimonious cladogram is a poor estimator of phylogeny.

Huelsenbeck's simulations were based on a model in which the probabilities of character change were equal along all branches of the cladogram. The accuracy of the most parsimonious cladogram resulting from these simulations is determined by the probability of character change along its branches. If the probability is very low, then most characters will be invariant or autapomorphic. If it is too great, then the distribution of the characters effectively becomes independent of the cladogram. Skewness is affected similarly. When characters are invariant or autapomorphic, skewness is 0. Skewness is weak when the probability of change is large and the expected frequencies of all states become equal, as in Hillis's null model. Skewness is strongest with intermediate probabilities. Accuracy of the most parsimonious cladogram is thus correlated with the degree of skewness of the DCL but that correlation is not general. It results simply from the constraint that character change has the same probability along each branch of the cladogram.

DCL skewness is also insensitive to the number of characters in the data set and thus does not necessarily reflect the degree to which conclusions are corroborated, a property that would be expected of a measure of phylogenetic structure in a data set.

Skewness as a measure of cladistic structure in data also suffers from two practical difficulties. First, informative cladistic data would be expected to show a strongly left-skewed DCL with a highly attenuated left tail. However, skewness is determined primarily by the central mass of the distribution. A negatively skewed DCL will occur whenever the median value exceeds the mean and thus need not necessarily have a long tail. Determining the degree of attenuation of the tail by estimating the lengths of all possible bifurcating cladograms is relatively quick for small numbers of taxa. However, as noted in Chapter 3, the number of such cladograms rises very quickly as the number of taxa increases, as a result of which sampling the cladograms becomes necessary in order to achieve a result in an acceptable time. The problem is that as the left tail becomes ever more attenuated, a sample of topologies is increasingly less likely to include cladograms from this region. Thus, the very cladograms upon which the significance test is based have little chance of affecting the skewness calculations. Second, the DCL is based only on fully bifurcating cladograms. If many of these cladograms represent arbitrary resolutions, then counting them as distinct can lead to erroneous conclusions.

To summarize, skewness of the DCL as a measure of the strength of cladistic structure in data consists of using a poorly chosen statistic to summarize a poorly chosen distribution.

6.2.3 Permutation tail probability (PTP)

It was noted above that, as a criterion for evaluating a cladogram in terms of its efficiency at summarizing the data, the ensemble consistency index (CI) suffers from the limitation that it is indifferent to the distribution of the states of characters among taxa. The ensemble retention index (RI) was specifically developed to remedy this deficiency by according greater influence to characters that support larger monophyletic groups and downweighting those that occur more distally on the cladogram. However, both CI and RI are expressed as proportions and are thus insensitive to the total number of characters, a deficiency shared with DCL skewness. Two data sets with identical CI and RI can differ in the amount of support they afford a cladogram if one data set includes a larger number of characters than the other.

These perceived limitations of CI and RI might be avoided if the number of characters and their distributions among taxa could both somehow be taken into account. CI measures the degree to which data are explained by a cladogram against a standard defined by the theoretical minimum number of steps possible for that data. This minimum can only be attained if the characters covary in such a way that they all fit the cladogram perfectly, that is, none of them shows any homoplasy. This cladistic covariation reflects the degree to which all of the characters are explainable by the same cladogram topology.

Faith and Cranston (1991) suggested that those most parsimonious clado-grams that entailed large amounts of homoplasy might derive from data sets in which the characters exhibited such poor cladistic covariation that analysis of a comparable set of randomly covarying characters could produce a cladogram of equal or even shorter length. If this were true, then doubt would be cast upon the validity of the original result. In order to determine whether the degree of character covariation in a data set is significantly greater than that expected for a set of comparable randomly varying charac-ters, Faith and Cranston proposed the permutation tail probability (PTP) test. The PTP test (Fig. 6.2) uses a randomization method in which the states of each character are permuted and randomly reallocated among the ingroup taxa in such a way that the proportions of each state are maintained. So, for example, given five ingroup taxa A–E, if a character occurs with state 0 in taxa A, B and C, and with state 1 in taxa D and E, permutation would maintain the 3:2 ratio of occurrence of state 0 to state 1, but might allocate state 0 to taxa A, C and E, and state 1 to taxa B and D. The character states of the outgroup taxa are held constant and not permuted. This procedure is applied to each character in the data set independently. In this way, the identity of each character and the frequencies of each of its states is maintained, but the cladistic covariation among the characters is disrupted. The procedure is then repeated, say 99 times, to give 99 new data sets, which are then analysed using standard parsimony techniques. The lengths of each of the most parsimonious cladograms from these 99 analyses are then compared to the length of those derived from the original matrix. The PTP is then defined as the proportion of all data sets (permuted plus original) that yield cladograms equal to or shorter than those produced from the original data set, and might be interpreted as the probability of obtaining a cladogram of this length under a model of random character covariation.

The value of the PTP has also been employed as the basis of a statistical test to assess the degree of cladistic structure in the data. The null hypothesis of this test is that there is no cladistic structure beyond that due to chance and, for example, would be rejected at the 0.05% level if the most parsimo-nious cladograms of fewer than five of the 100 data sets were as short or shorter than those derived from the unpermuted data (PTP ≤ 0.05). When none of the permuted data sets produces minimum length cladograms as short as those derived from the real data, then Faith and Cranston suggested

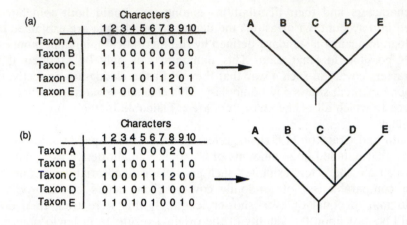

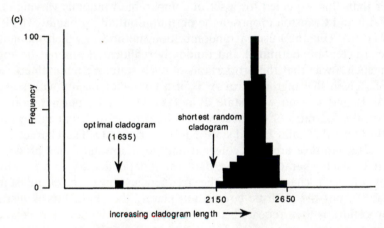

Fig. 6.2 PTP analysis. (a) The most parsimonious cladogram for the original data set is determined and its length recorded. (b) The states of each character are then permuted among the taxa, while maintaining the proportions of each state, to produce a new data set. This data set is then analysed and the length of its most parsimonious cladogram recorded. (c) This procedure is then repeated a large number of times and the PTP is defined as the proportion of all data sets (permuted plus original) that yield cladograms equal to or shorter than those produced from the original data set. The null hypothesis that there is no cladistic structure beyond that due to chance would be rejected at the 0.05% level if the most parsimonious cladograms of fewer than 5 of the 100 data sets were as short or shorter than those derived from the unpermuted data (PTP ≤ 0.05).

that the PTP be arbitrarily recorded as 0.01. Thus a low PTP is desirable and a value of less than 0.05 could be taken to imply the presence of significant cladistic covariation or structure in the original data.

However, because the PTP test is based upon the determination of the lengths of most parsimonious cladograms, there is a practical problem with its application. For smaller data sets, it may be possible to apply exact methods of cladogram construction and thus obtain precise estimates of the PTP value in a reasonable computational time. However, this is not possible for larger data sets, which must be analysed using heuristic methods that are not guaranteed to find the minimum length solutions. Furthermore, the significance level, α, of a PTP test cannot be less than $1/(W+1)$, where W is the number of permutations. For example, if a more stringent α value of 0.01 is required, then 99 permutations must be undertaken, which may result in excessively long computational times. Faith and Cranston noted this drawback and suggested that the PTP could perhaps be estimated by applying the same heuristic procedure to both types of data set, i.e. original and permuted. Alternatively, it might be sufficient simply to compare the results from an exhaustive search of the real data with those of heuristic searches of the permuted data. However, Källersjö *et al.* (1992) noted that, because such approximate values would always exceed the corresponding exact values, this apparently conservative latter procedure would actually increase the apparent difference between the real and randomized data sets and worsen the risk of a false conclusion of significant congruence.

Instead, they suggested using the single pass 'hennig' command of the Hennig86 program (Farris 1988) to estimate very quickly the length of the most parsimonious cladogram. This approximate length seldom differs by more than a few per cent from that of the most parsimonious cladogram calculated by an exact method. The test would then use only the number of permutations in which the lengths of the estimated minimum length cladograms exceeded that for the real data. If this procedure is applied to both types of data and repeated many times (say 10 000), then the approximation differences would be unlikely to have much effect on that number. The successful results of such an application should be similar to the histogram depicted in Fig. 6.1. The bar furthest to the left represents the length of the estimated minimum length cladogram obtained from the real data. This is well separated from the rest of the distribution and thus there is little chance that the use of approximate methods will lead to an erroneous conclusion.

In addition to this practical problem, the validity of the PTP test has also been questioned from a theoretical viewpoint. Bryant (1992) argued that the null hypothesis of randomly covarying characters is contrary to the very basis of cladistics. Every character is a putative synapomorphy or homology statement, from which it follows axiomatically that cladistic characters will covary hierarchically. Characters are thus intrinsically hierarchical and cladistic analysis simply summarizes the total implied hierarchy in the data set as efficiently as possible. It is the expected covariation among characters that is the assumption that justifies the search for the most parsimonious cladogram. Furthermore, because hierarchy is implicit in the data, it must follow that cladograms are inherently hierarchical. The two are not independent. In contrast, permuted data, in which the covariation among characters has been deliberately disrupted, can have no intrinsic hierarchy. When analysed by cladistic methods, such data must have hierarchy imposed upon them in order to produce a cladogram. Thus the PTP test compares the lengths of real cladograms, which have inherent hierarchy, with those of contrived cladograms upon which hierarchical order has been imposed.

Carpenter (1992) considered that Faith and Cranston had failed to provide any justification why their particular randomization procedure or significance test should be chosen to specify probabilities. Why not simply allocate the entries in the data matrix completely randomly, with no constraints as to the values that could be obtained? Carpenter thus considered that the PTP test, and permutation procedures in general, amounted to 'no more than a misapplication of statistics, and add nothing to the use of cladistic parsimony'.

Hence, rather than providing a criterion for the acceptance or rejection of a cladogram, the results of a PTP test might be better viewed as a relative measure of overall confidence in the data set and may provide independent evaluation of the explanatory power of the data. If the length of the most parsimonious cladogram derived from real data, which is inherently hierarchical and covariant, is not significantly different from the length of the shortest cladogram derived from permuted data, which lacks covariation and upon which hierarchy must be imposed, then the explanatory power of the real data set is low. Conversely, if the most parsimonious cladogram from the real data is considerably shorter than those derived from any permuted data set, then we can have increased confidence in our result. However, while the PTP test can indicate that a data set contains significant cladistic structure, the converse does not pertain. A most parsimonious cladogram obtained from a real data set that is similar in length to those obtained from permuted data does not denote that this data lacks cladistic structure. It merely reduces our confidence in the overall support for that pattern. The most parsimonious cladogram remains the best estimate of the relationships among the study taxa based upon the data to hand.

6.3 SUPPORT FOR INDIVIDUAL CLADES ON A CLADOGRAM

In Chapter 1, it was explained that the recognition of monophyletic groups is predicated upon the discovery of synapomorphic characters. In theory, only one synapomorphy is required to establish the validity of a monophyletic group. However, as the number of independent synapomorphies supporting a clade increases, then so does both its corroboration and our confidence in it as a hypothesis of relationships. Therefore, the most straightforward assessment of the support that can be accorded a clade is simply to count the number of characters on the branch subtending it. In other words, branch length equates to branch (clade) support.

However, this equation is an idealized concept that can only be applied objectively and unequivocally when all characters on a cladogram are unique and unreversed synapomorphies. Homoplasy makes the assessment of branch support difficult, as attempts are made, for example, to weigh a reversal in one character against an independent development in another. The results of such considerations are often highly subjective, even authoritarian. In addition, homoplasy need not be evenly distributed over the cladogram but concentrated on certain branches. Such branches may thus appear to be much better supported than they actually are if branch length alone is used as the criterion of support. In the presence of homoplasy, branch length is merely a conjecture of support, not an absolute measure, and may be misleading.

6.3.1 Bremer support

One method that has been proposed as a more precise measure of clade support is the number of extra steps required before a clade is lost from the strict consensus tree of near-minimum-length cladograms. This notion has been variously referred to as 'Bremer support' (Källersjö *et al.* 1992), 'branch support' (Bremer 1994), 'length difference' (Faith 1991), 'clade stability', '[Bremer's] support index' (both Davis 1993) and 'decay index' (Donoghue *et al.* 1992). This last term is perhaps unfortunate, if only because the best supported groups are those with the greatest decay. As all of these terms are synonymous, we choose Bremer support, both because it is unambiguous and in recognition of the author who first applied the concept in the context of parsimony analysis (Bremer 1988).

To calculate Bremer support (Fig. 6.3) for a particular clade of the most parsimonious cladogram, all the cladograms one step longer than the minimum are found. Then, the strict consensus of these plus the most parsimonious cladogram is constructed. This process is repeated, increasing the length of the suboptimal cladograms by one step each time until the clade in question no longer occurs on the consensus. The number of extra steps required to reach this point is the Bremer support for the clade. When only a single most parsimonious cladogram is found for a data set, all the included clades must have Bremer support >1. Should there be more than one equally parsimonious cladogram, then Bremer support calculation begins with the construction of either the strict or the semi-strict consensus tree of these cladograms (see Chapter 7 for details of consensus methods). Some groups will necessarily be lost on this first consensus (otherwise there would not have been any alternative topologies in the first place) and such groups will have zero Bremer support. Thus, whenever there are multiple equally parsimonious solutions, at least one clade will have zero Bremer support.

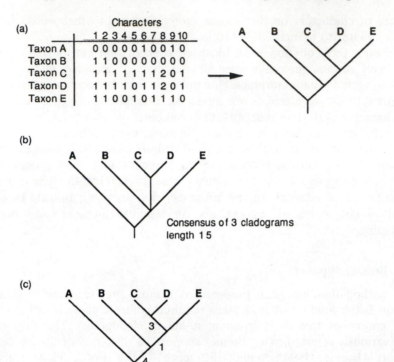

Fig. 6.3 Bremer support. (a) The most parsimonious cladogram is determined for a data set. (b) All cladograms that are one step longer than the most parsimonious are then determined. The strict consensus tree of these plus the most parsimonious cladogram is constructed to determine those groups that no longer receive unambiguous support. (c) This process is repeated, increasing the length of the suboptimal cladograms by one step each time, until all groups are lost. The Bremer support for a group is the number of steps that have to be added before that group is no longer recovered on the strict consensus tree of optimal and suboptimal cladograms.

Bremer support has advantages over simple branch length as a measure of strength of evidence for a clade. When all characters are perfectly congruent, the most parsimonious reconstruction is unique and Bremer support equates to branch length. Otherwise, the support for a clade is reduced to the degree that there are alternative equally parsimonious groupings or character optimizations.

Bremer support values have been used as the basis of two methods to assess the overall support accorded to a cladogram. 'Total support' is simply the sum of all the Bremer support values over the cladogram. Total support can be rescaled meaningfully because its upper bound is the length of the most parsimonious cladogram. This follows from the fact that the support for a branch cannot exceed its length. The 'total support index', ti, is thus defined as the ratio of total support to the length of the most parsimonious clado-gram. When the strict consensus tree of the most parsimonious cladograms is completely unresolved, then all groups have zero Bremer support and ti = 0. In contrast, if there is no homoplasy in the data, and hence only a single most parsimonious cladogram that is fully supported by all characters, then ti = 1. The total support index measures the stability of the most parsimonious cladogram(s) in terms of supported resolution, rather than in terms of the degree of homoplasy in the data. However, a low ti value should not be taken to imply that all the clades on a cladogram are poorly supported. Some clades may have high Bremer support values, despite low total support.

6.3.2 Randomization procedures

An alternative and profoundly different approach to the assessment of support for individual clades applies data perturbation and randomization. Three basic types of randomization methods can be recognized: permutation, jackknifing and Monte Carlo. Jackknifing can be further divided into simple first-order jackknifing and more complex, higher-order jackknifing. Monte Carlo methods can be divided into those that are model-dependent and those that are model-independent. Of the latter, the best known and most fre-quently used is the bootstrap.

Bootstrap
Applied to cladistic data, the bootstrap (Fig. 6.4) randomly samples characters with replacement to form a pseudoreplicate data set of the same dimensions as the original. The effect is to delete some characters randomly and to reweight others randomly, with the constraint that the sum of the weights for all characters equals the number of characters in the matrix. A large number of pseudoreplicates is generated, typically 1000 or more. The most parsimo-nious cladograms for each pseudoreplicate are then found and the degree of conflict among them assessed by means of a majority rule consensus tree, which includes all those groupings that are supported by more than 50% of the pseudoreplicates (see Chapter 7 for further details). The percentage of most parsimonious cladograms resulting from the pseudoreplicates in which a particular group is found might be interpreted as a confidence level associ-ated with that group. For example, if a group appears in 95% or more of these cladograms, then it could be concluded that this group is supported at the 95% level.

However, there are several serious limitations to this use of the bootstrap. First, for such confidence limits to be valid, the groups for which the monophyly is to be tested should be specified in advance. If we cannot specify any such groups, then the number of potential groups is so large that in order to maintain an overall type I error rate of say 0.05, such an exceedingly low α level would be required that the final confidence interval would be vastly

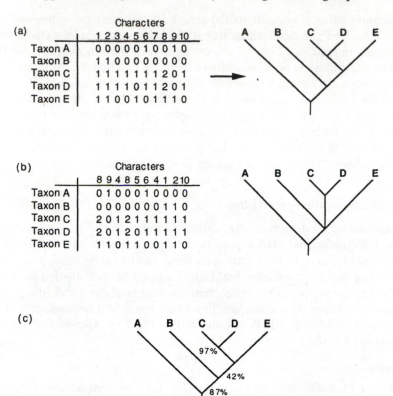

Fig. 6.4 Bootstrap analysis. (a) The most parsimonious cladogram is determined for the original data set. (b) Characters are then sampled randomly with replacement to produce a pseudoreplicate data set of the same size as the original. Some characters (e.g. #8) will be represented more than once, while others (e.g. #3) will not be included at all. The most parsimonious cladograms for the pseudoreplicate data set are then constructed. (c) This process is repeated a large number of times (e.g. 1000) and the results summarized by means of a majority-rule consensus tree. Support for a group is then interpreted as the percentage of most parsimonious cladograms resulting from the pseudoreplicates in which the group is found.

inclusive (Swofford and Olsen 1990). Second, the confidence intervals obtained through resampling methods are only approximate unless the original sample size, that is, the number of characters in the data matrix, is large. This is 'large' in the statistical sense (more than 1000 and preferably 10 000) and most data sets do not begin to approach this number of characters. Even molecular data, which can comprise several thousand base pairs, do not contain this number of informative sites. It has been argued that bootstrapping is not affected by the inclusion of uninformative characters (Harshman 1994) but this has been shown to be incorrect (Carpenter 1996). Thus, few, if any, data sets meet the statistical size requirement of the bootstrap.

An alternative view of the bootstrap is that it indicates how the support for the various groups on the most parsimonious cladogram is distributed among characters. The expectation is that clades supported by a large number of characters will be recovered frequently and receive high scores on a majority rule consensus tree. In contrast, clades that are supported by only a few characters, especially if these are homoplastic, are not expected to be recovered very often, if at all. However, a clade may be unambiguously supported by a single character on the most parsimonious cladogram yet fail to be recovered by a bootstrap analysis, due to the random nature of the resampling. Thus, groups can be excluded from the majority rule consensus tree even though they are uncontradicted on the most parsimonious clado-gram. The bootstrap thus provides only a one-sided test of a cladogram. Groups that are recovered are supported by the data, but groups that are not recovered cannot be taken as rejected.

There is, however, a more serious and fundamental problem with the bootstrap. This is the requirement that the characters in the original data matrix should represent a random sample of all possible characters. However, in systematic studies, characters are not randomly sampled from independent, identically distributed populations. Rather, they are carefully selected and filtered with the aim of best resolving the relationships of the taxa under study. Such systematic bias is not considered to be a problem by advocates of the bootstrap, who assert that attempts by systematists to try to ensure that their characters are independent and uncorrelated are sufficient. However, subjectively trying to ensure that characters are independent of one another is simply inadequate. Unless the characters in a data set do accurately reflect the larger underlying distribution of all possible characters, then bootstrap confidence intervals may be very poorly estimated. There are also many other factors that might lead to either overestimates or underestimates of confidence, including size of the data set, efficiency of any heuristic search procedures that are employed, cladogram topology, and differential and uneven rates of character change among branches.

Thus, at best, the effects of these limitations mean that we would be unwise to treat bootstrap confidence intervals as absolutes, although they may serve as approximate guides to the support afforded to groups by the data. At worst, the application of the bootstrap in cladistic studies can be considered to lack rigorous justification.

Jackknife

In contrast to the bootstrap, jackknife sampling is applied without replace-ment and hence the pseudoreplicate data sets are smaller than the original. Jackknifing aims to achieve better variance estimates than might otherwise be possible from small samples. In first-order jackknifing, pseudoreplicates are constructed by randomly removing one observation (taxon or character) from the data set. Hence, for a data set of T observations, T pseudoreplicates are possible, each comprising $T-1$ observations from the original sample. The variances of the T pseudoreplicates are then averaged to give the estimate of the parametric variance.

First-order jackknifing of taxa was introduced into systematics by Lanyon (1985). If a data set contains no homoplasy, then deletion of one taxon will have no effect on the topology of the most parsimonious cladogram. This is because the information contained in the synapomorphies of that taxon is inherent in those taxa that occupy more distal positions on the cladogram. The most parsimonious cladogram obtained from analysis of a sample of $T-1$ taxa will thus be identical to that obtained from analysis of the complete data set, but with the terminal branch leading to the deleted taxon pruned out. However, if there is conflict in the data, then analysis of jackknifed data sets need not produce the same topology as that derived from the complete data set. Any conflict is revealed by constructing a strict consensus tree of all the most parsimonious cladograms derived from all possible first-order pseudoreplicates by means of a strict consensus tree (that is, a consensus tree that contains only those components common to all of the fundamental cladograms; see Chapter 7 for further explanation). However, the normal strict consensus method discards those taxa not present in all the fundamental cladograms and thus applying this procedure to cladograms produced from jackknife pseudoreplicates would produce a consensus tree that contained only the outgroup. Lanyon (1985) proposed a modification to allow for the deleted taxa, called the jackknife strict consensus tree, which contains those nodes that are shared by or consistent with all of the pseudoreplicate cladograms.

However, Lanyon's advocacy of a strict consensus tree suffers from the drawback that strict consensus is indifferent to the proportion of most parsimonious cladograms in which a clade is supported. Thus, it takes only one cladogram from one pseudoreplicate that wildly disagrees with the remainder in the position of one taxon to collapse the entire consensus into a bush. To circumvent this problem, Siddall (1996) proposed the jackknife monophyly index (JMI), which is defined as:

$$\text{JMI}_c = \frac{\sum\limits_{t=1}^{T} \rho(c_t)}{T}$$

where T is the number of ingroup taxa and $\rho(c_t)$ is the proportion of the most parsimonious cladograms of pseudoreplicate t in which clade c is supported. Because the JMI is monophyly-dependent (i.e. it is calculated using rooted cladograms), it is inappropriate to jackknife the outgroup taxa (analogous arguments have been put forward with regard to the PTP test). Siddall advised against using JMI values as the basis for accepting or rejecting individual clades. While the JMI might, in some sense, indicate the amount of statistical support there is for a clade, it certainly cannot be used to argue the converse, that is, the degree of support against a clade. The JMI does show which groupings on a cladogram are more stable and which are less stable. The calculations can also help to identify 'critical' and 'problematic' taxa. Critical taxa are those whose deletion results in a great increase in the number of most parsimonious cladograms. In contrast, the deletion of problematic taxa stabilizes the results and reduces the number of most parsimonious cladograms.

Higher-order jackknifing removes subsets of n observations to give pseudoreplicates of size $T - n$. As there is no justification for stopping at a particular subset size, the removal of all possible subset sizes from 2 to $(T - 1)$ should be investigated. However, the number of possible ways in which any number of taxa (up to $T - 1$) can be removed from T taxa is $2^T - 2$. For more than about ten taxa, this results in an impractically large number of analyses to perform. Random sampling of all possible combinations might provide a suitable heuristic solution, which could be performed by randomly choosing a subset size and then randomly deleting this number of taxa. This procedure is identical to bootstrapping taxa, because this latter method randomly deletes some taxa and randomly replicates others. As the latter are duplicates identical in character information, they would act as one terminal taxon in a parsimony analysis. The overall effect would thus be the same as randomly deleting a random number of taxa, i.e. a higher-order jackknife. Siddall (1996) rather irreverently called this method the 'jackboot'. Application of the jackboot would only be appropriate if the observations being jackknifed could be assumed to be drawn randomly from some larger sample, an assumption that might prove difficult to justify for taxa. Furthermore, the effects of higher-order subset deletions might be expected to be more severe than those of lower-order subsets. How these effects would be weighted differentially remains unclear.

First-order jackknifing of characters was first applied in a systematic context by Mueller & Ayala (1982), who advocated it for estimating the sampling variance of Nei's genetic distance. However, first-order jackknifing is of limited use when applied to cladistic characters because it amounts to little more than asking if there is more than one apomorphy supporting a clade. In a homoplasy-free cladogram, all clades with only a single supporting apomorphy would not be recovered, while all other clades would receive a perfect score regardless of the number of apomorphies supporting them. Likewise, higher-order jackknifing simply extends this problem and ultimately, only those clades supported by at least as many characters as there are taxa are guaranteed to be recovered.

Clade stability index

Davis (1993) proposed a method for estimating the stability of clades that is similar to higher-order jackknifing. However, rather than using character deletion as a means of estimating sample variance, Davis used it to identify those characters or combinations of characters that are critical to the maintenance of clades on a cladogram. Characters are successively deleted, first individually then as increasingly larger subsets until the clade under study is lost from the strict consensus tree of most parsimonious cladograms resulting from the analysis of the reduced data sets. The clade stability index (CSI) is defined as the ratio of the minimum number of character deletions required to lose a clade to the total number of informative characters in the data set. Thus, a clade that is lost after the removal of two characters from an informative data set of 15 characters would receive a CSI of $2/15 = 0.13$.

The CSI suffers from the same computational difficulties as higher-order jackknifing as the number of characters increases. For large data sets, Davis suggested it would be sufficient to analyse all reduced data sets with one and all combinations of two characters removed, together with 500 reduced data sets each for subsets of three to ten characters. For his data set of 74 characters, this would entail analysing 6775 data sets. Under such a strategy, the CSI would be a maximum value because an unsampled combination of character deletions may exist that would prevent a clade from being resolved.

The value of the CSI is also strongly influenced by the size of the data set. For a given amount of apomorphic support for a clade, the larger the data set, the smaller the CSI, a deficiency that calls into question the suitability of the CSI as a general measure of clade support. Furthermore, expressing clade support in terms of characters works only with binary and non-additively coded multistate characters. For additive binary characters, it is possible for a clade to be supported by more than one character state change in a given character. Deletion of a single character may then lead to the loss of more than one apomorphic change supporting a clade. For these reasons, and in addition being easier to calculate, Bremer support is to be preferred over the CSI as a means of estimating clade support.

Topology-dependent permutation tail probability (T-PTP)

The permutation methods of the PTP test have been extended (Faith 1991) to provide both *a priori* and *a posteriori* tests of the monophyly or non-monophyly of a selected clade on a cladogram: the 'topology-dependent cladistic permutation tail probability' (T-PTP) tests. Evidence for the mono-phyly of a clade is considered strong if the most parsimonious cladogram in which the clade is monophyletic is much shorter than the most parsimonious cladogram obtainable under the constraint that the clade is not monophyletic (i.e. paraphyletic or polyphyletic). The strength of the evidence is assessed by comparing the length difference under the two topological constraints for real data against that obtained from permuted data. As in the PTP test, the significance value is calculated as the proportion of occasions on which the observed length difference is equalled or exceeded by the value from the permuted data.

A priori application of the T-PTP test requires that the monophyletic group to be tested be hypothesized before the cladistic analysis is carried out and asks whether a level of support for this group can be found that represents a significant departure from randomness. In contrast, the *a posteriori* T-PTP test asks whether the support for a given clade found as the result of a cladistic analysis could have arisen by chance. The test parameter is the same as that in the *a priori* T-PTP test, that is, the length difference between the most parsimonious cladograms obtainable under the constraints of mono-phyly and non-monophyly. However, for each permuted data set, the length difference is now calculated as the largest value that could be achieved for any monophyletic group of the same size as the clade under consideration. The T-PTP value is then estimated as the proportion of all data sets (real plus permuted) in which some monophyletic group can be found with a difference value at least as large as that found for the clade in question.

T-PTP can be extended to include tests for the non-monophyly of a group or, indeed, any conceivable set of compatible topological constraints. One such set comprises all the clades of a given cladogram and is referred to as the 'all-groups' form of the T-PTP, or T-PTP(AG). In this version, the test parameter is the difference in length between the most parsimonious clado-gram and the minimum length cladogram obtainable that includes none of the clades of the original topology. As such, it is actually a measure of support for an entire cladogram, rather than individual components thereof.

However, as a derivative of the PTP test, T-PTP tests suffer from all the theoretical and practical defects of that test. Hence, an *a posteriori* T-PTP test cannot provide a criterion for the acceptance or rejection of a clade, but only give a relative measure of confidence in that clade.

6.4 SUMMARY

In summary, no method proposed so far can serve as more than a useful general guide to the degree of confidence that we may place in the results of a cladistic analysis. Certainly, none of the statistics discussed above can place statistically meaningful confidence limits on to cladograms. Furthermore, if characters are intrinsically hierarchical, then any method that uses a random model as the basis for a null hypothesis is entirely inappropriate. However, while the most parsimonious cladogram remains the best estimate of the relationships among the study taxa based upon the data to hand, there is a need to develop rigorous measures of confidence appropriate to the clado-grams, in order that we avoid overinterpreting the significance of our results. But these measures of confidence in a group should be based upon the probability that the characters considered to be synapomorphies have been correctly interpreted (Farris, in Werdelin 1989).

6.5 CHAPTER SUMMARY

1. Many methods have been proposed that can purportedly be used to assign degrees of support or confidence to the results of cladistic analysis. These can be divided into two types. The first are applied to the cladogram as a whole and aim to answer the question of whether the data contain 'significant' cladistic structure and thus the resultant cladogram is not just the product of chance. The second group of methods examine the support afforded to individual groups on a cladogram, in order to determine which are well supported by data and which are only weakly supported.

2. Methods aimed at assessing the support of whole cladograms all use the same general principle. The real data set is repeatedly perturbed according to a set of rules to produce a large number of data sets of 'phylogenetically uninformative' data. The lengths of the most parsimonious cladograms derived from these contrived data sets are then compared with the length of the most parsimonious cladograms for the real data set, with the expectation that the latter will be substantially shorter than any of the former. These methods include data decisiveness (DD), distribution of cladogram lengths (DCL), and the permutation tail probability (PTP) test.

3. DD compares the real data with a matrix of completely undecisive data, that is, one in which all possible informative characters occur in equal numbers. If the DD value is small, then there are only very weak reasons for preferring the most parsimonious cladogram derived from the real data. In contrast, a high DD value does not signify high information content and certainly cannot be used to assign a confidence level to the most parsimonious cladogram for the real data.

4. The use of DCL as an indicator of phylogenetic signal is based on the premise that when the DCL is nearly symmetrical, many cladograms will be only a few steps longer than the most parsimonious cladogram and thus the phylogenetic signal is weak. However, if the DCL is strongly negatively or left-skewed, there will be very few cladograms that are just slightly longer than the most parsimonious solution and thus the phylogenetic signal is strong. However, DCL skewness is a poor measure of phylogenetic support because it is strongly dependent upon the probabilities of character change along each branch of a cladogram and is also insensitive to the number of characters (and thus to the degree to which conclusions are corroborated).

5. In the PTP test, the data are perturbed in such a way as to break any cladistic covariation among the characters. This is achieved by randomly reallocating states within characters among taxa while maintaining the

original proportions of each state. Characters are treated independently and the states in the outgroup taxa are held constant and not permuted. However, the null hypothesis of randomly covarying characters is contrary to the basis of cladistics. Cladistic characters are intrinsically hierarchical and cladistic analysis aims to summarize this hierarchy as efficiently as possible.

6. The simplest measure of support for individual clades is branch length. However, homoplasy makes the objective interpretation of branch length as support difficult. Bremer support aims to circumvent this problem by assessing the number of extra steps that are required before a clade is lost from the strict consensus tree of near-minimum length cladograms. When there is no homoplasy, Bremer support is equal to branch length, otherwise the support for a clade is reduced to the degree that there are alternative equally parsimonious groups or character optimizations.

7. There are three basic types of approaches that use randomization procedures to assess support for individual clades: Monte Carlo methods (including bootstrapping), jackknifing and permutation procedures.

8. Monte Carlo methods can be model-dependent or model-independent. Of the latter, the best known is the bootstrap. A large number of pseudoreplicate data sets of the same size as the original are created by randomly sampling characters with replacement. The effect is to delete some characters randomly and to reweight the rest randomly. The most parsimonious cladograms for these pseudoreplicates are calculated and a majority rule consensus tree used to assess the degree of conflict among them. The percentage of pseudoreplicate data sets that recover a given group is then interpreted as a measure of confidence in that group. However, the bootstrap makes several assumptions that severely limit its usefulness in this regard. The most serious of these is that the characters in the data set are a random sample from all possible characters. In practice, systematists never sample characters at random but carefully select them, thus violating the theoretical underpinning of the bootstrap as a measure of clade confidence.

9. The jackknife creates pseudoreplicate data sets that are smaller than the original by sampling without replacement. First-order jackknifing removes only one character or taxon; higher-order jackknifing removes two or more. Jackknifing of taxa can help us to identify 'critical' and 'problematic' taxa. Deletion of critical taxa results in a great increase in the number of equally most parsimonious solutions, while the deletion of problematic taxa has the converse effect.

10. The clade stability index is defined as the ratio of the minimum number of character deletions required to lose a clade on a cladogram to the total

number of informative characters in the data set. However, it suffers from major computational difficulties and the deficiency that for a given amount of apomorphic support for a clade, the larger the data set, the smaller the CSI. For these reasons, the use of Bremer support is preferred to the CSI.

11. The PTP test is applied to individual clades as the 'topology-dependent cladistic permutation tail probability' (T-PTP) test. However, as a derivative of the PTP test, the T-PTP test suffers from all the same flaws and failings and is of very limited utility.

12. None of the methods advanced so far can place meaningful confidence limits onto cladograms, although data decisiveness and Bremer support may serve as useful general guides. If cladistic characters are intrinsically hierarchical, then all methods using a random model as a null hypothesis are fundamentally flawed.

7.

Consensus trees

7.1 INTRODUCTION

In a cladistic analysis of a particular group of taxa, data are frequently obtained from several sources. In biogeographical and co-evolutionary studies, information may be available for two or more groups of taxa. Systematists generally hold that because different character sets share common evolutionary histories, then reliable phylogenetic methods should be able to recover a common resolved pattern. This reasoning led to the idea that congruence among data sets and fundamental cladograms might provide the strongest evidence that phylogenetic reconstruction is accurate (Penny and Hendy 1986, Swofford 1991). In practice, however, analysis of different data sets can produce cladograms with everything from minor discrepancies in the placement of one taxon to completely different overall topologies.

When confronted with differences between two or more cladograms, and especially when they are very disparate, we are faced with a number of problems. For example, for different data sets and the same group of organisms, incongruence can be explained by suggesting that at least one of the data sets is wrong or that each data set is providing only part of the 'correct' systematic signal. The issues of debate concern the relative strengths of particular data sets, particularly of morphological and molecular data sets, and the reliability or futility of different analytical methods. However, the underlying issue remains whether there any methods that can be used to differentiate unavoidable incongruence from 'spurious' differences due to sampling error.

This chapter reviews the more accessible consensus methods currently used to determine congruence and incongruence among different data sets, but avoids the issue of 'total evidence' versus 'consensus' (reviewed in Chapter 8). Since the advent of computer packages for cladistic analysis, there have been considerable developments both in methods of cladogram comparison and in statistics for assessing levels of agreement in seemingly very different data sets and topologies.

Consensus methods are a convenient means of summarizing agreement and disagreement, or congruence and incongruence, between two or more cladograms. Common to all methods of consensus analysis is the desire to construct a tree from the non-contradictory components found among the set of cladograms generated from the initial analysis. Consensus trees can be considered to be indirect methods for resolving character conflict in the construction of a general classification. They reduce the number of fundamental cladograms produced by parsimony analysis to one tree showing their common components (i.e. non-trivial or informative groups *sensu* Nelson and Platnick 1981).

Cladograms generated from a data set in a cladistic analysis are called fundamental because they summarize its hierarchical information. In contrast, consensus trees are derivative, being constructed from and representing a set, or sets, of cladograms. Consequently, consensus analysis almost invariably produces a tree that would not be supported as most parsimonious by the original data and in some cases may even contain components not found in any of the fundamental cladograms. For this reason, some authors (e.g. Miyamoto 1985, Carpenter 1988) have argued against their use on the grounds that a consensus tree rarely summarizes a data set as efficiently as any one of the fundamental cladograms from which it was constructed. Nevertheless, consensus trees have their uses in allowing investigation of data concordance amongst cladograms generated from different data sets. This is important for investigating 'difficult' taxa that occur in different positions and thus produce fundamental cladograms with different topologies. It is important also for finding intersections amongst cladograms when comparing 'hosts' and 'associates' (Page 1993*b*), for example, between gene trees and species cladograms, hosts and their parasites in co-evolutionary studies, and area cladograms and biological cladograms in historical biogeography (Humphries and Parenti in press). Others argue that because cladistic analyses of almost all data sets produce multiple equally most parsimonious cladograms, then consensus trees should be considered as essential in classification because they provide the only reasonable summary of the information that can be achieved (Anderberg and Tehler 1990, Bremer 1990).

Consensus analysis has become a minor growth industry in systematics and many methods have been described. This chapter describes only the seven most commonly encountered in the cladistic literature: strict, majority-rule, combinable components (or semi-strict), Nelson, Adams, agreement subtrees (or common pruned trees), and median consensus trees.

Strict and majority-rule consensus methods are based on simple counts of the frequency of informative groups (components) in the set of cladograms being compared. Strict consensus trees contain only those components common to all of the fundamental cladograms. Combinable components or semi-strict consensus trees (Bremer 1990) include all of the components found in strict trees but in addition include those components that are uncontradicted by less resolved components within the set of fundamental cladograms. Majority-rule consensus trees contain all of those clusters found in more than 50% of these cladograms. The cut-off value can be set to a higher value, say 75%, but then the result is not strictly a *majority-rule* consensus tree. Median consensus trees are closely related to majority-rule consensus trees. This method uses a metric that measures topological disagreement between any pair of cladograms (Barthélemy and Monjardet 1981, Barthélemy and McMorris 1986). Nelson consensus trees comprise the cliques of mutually compatible components that are most replicated in the fundamental cladograms. Adams trees contain all intersecting sets of taxa common to all cladograms. In the greatest agreement subtree method, the least number of branches are 'pruned' from the fundamental cladograms to produce the largest subtree with greatest agreement. In the trivial comparison of two cladograms, strict and majority-rule consensus trees are identical, as are combinable components and Nelson consensus trees.

Nixon and Carpenter (1996*b*) argued that if the goal of consensus analysis is to summarize agreement in grouping among a set of fundamental cladograms, then only the strict consensus tree fulfils that goal. All the other methods listed above may yield trees with groups that are not fully supported by all the data or are supported only ambiguously. These Nixon and Carpenter called 'compromise trees', reserving the term consensus tree for the strict consensus method only.

However, while we recognize the fundamental distinction between strict consensus and other methods, we consider that there are equally fundamental differences among the other methods, and that the dichotomy proposed by Nixon and Carpenter is unnecessarily doctrinaire. We therefore retain the term consensus for all methods that aim to summarize the common information contained in a set of cladograms according to some specific criterion. Different consensus methods are suited to different tasks, although the literature is bewildering in that methods are frequently applied inappropriately and the terminology has become somewhat confused (Nixon and Carpenter 1996). Strict consensus is useful for determining common components of all fundamental cladograms, while combinable components consensus highlights resolved components amongst a profile of cladograms some of which contain unresolved components. Majority-rule consensus is most frequently used to summarize the results of bootstrap analyses (see Chapter 6). Adams consensus trees are used mostly to determine the degree of preserved structure in cladograms. They have their greatest value in comparing seemingly different topologies due to the erratic performance of 'rogue' taxa that appear in widely different positions on cladograms. Largest common pruned trees can be useful for determining incongruence in cladograms when only a few taxa are responsible for the different topologies.

7.2 STRICT CONSENSUS TREES

The most conservative consensus tree is a strict consensus tree. First used by Schuh and Polhemus (1981), the 'strict tree' was defined by Sokal and Rohlf (1981) as the unique tree that contains only those groups that occur in *all*

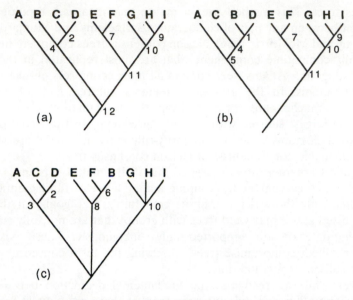

Fig. 7.1 Cladograms of (a) butterflies, (b) birds and (c) bats, calculated using COMPONENT 2.0 for Windows®. (From Bremer 1990.) Numbers refer to components in Table 7.1.

rival cladograms. These were also called 'Nelson consensus trees' by Schuh and Farris (1981). As Nelson's (1979) consensus method of adding together replicating and non-replicating components is different from strict consensus, the distinction made by Page (1989) is maintained here (see below).

A strict consensus tree is derived by combining only those components that appear in all members of a set of fundamental cladograms. Consider three cladograms for butterflies, birds and bats (Fig. 7.1). In a fully resolved cladogram, there are $n - 2$ informative components. Thus, for a cladogram of 9 taxa, there are 7 informative components. However, among the three cladograms in Fig. 7.1, we actually find 12 different informative components (Table 7.1), indicating that there is conflict among the cladograms. Of these 12, only component 10, comprising the group GHI, occurs in all three fundamental cladograms. The strict consensus tree thus contains only this component (Fig. 7.2a). The rationale behind strict consensus is that the data are only consistent for this one component. In this particular example, the conservativeness of strict consensus means that one is left with depressingly little resolution, although this is not always the case.

We noted in Chapter 4 that both PAUP and Hennig86 can generate spurious cladograms that are due solely to ambiguous character optimization. The length of the strict consensus tree can be used to determine whether all the apparent resolution found in a set of fundamental cladograms is due to such ambiguity. Suppose we have two cladograms that contain two fully

Table 7.1 Components of the butterflies, birds and bats cladograms, analysed using COMPONENT 2.0 for Windows® (Page 1993b). Rows (integers) are the 12 components found in these three cladograms; columns (letters) are the taxa, as labelled in Fig. 7.1. The composition of each component is indicated by the asterisks and the number of cladograms in which it occurs is given in the last column. For example, component 10, comprising taxa G, H and I, is present in all three of the cladograms, while component 8, comprising taxa B, E and F, occurs in only a single cladogram (bats). See text for further explanation.

	A	B	C	D	E	F	G	H	I	Occurrences
1	.	*	.	*	.	.	.	.	.	1
2	.	.	*	*	.	.	.	.	.	2
3	*	.	*	*	.	.	.	.	.	1
4	.	*	*	*	.	.	.	.	.	2
5	*	*	*	*	.	.	.	.	.	1
6	.	*	.	.	.	*	.	.	.	1
7	.	.	.	.	*	*	.	.	.	2
8	.	*	.	.	*	*	.	.	.	1
9	.	.	.	.	.	.	.	*	*	2
10	.	.	.	.	.	.	*	*	*	3
11	.	.	.	.	*	*	*	*	*	2
12	.	*	*	*	*	*	*	*	*	1

resolved but conflicting groups, each unambiguously supported by data. Because the two groups are not common to both cladograms, neither will appear in the strict consensus tree. Instead, their component taxa will form a polytomy and the strict consensus tree will be longer than the two fundamental cladograms, because multiple origins of the characters supporting each group will have to be postulated. However, if the difference in topology between the two conflicting groups is due to ambiguous character optimization, then collapsing the zero-length branches will have no effect on length, and the strict consensus tree will be the same length as the fundamental cladograms. In other words, if we find that a strict consensus tree is the same length as its fundamental cladograms, then we can conclude that all the extra resolution present in those cladograms is spurious and due to ambiguous character optimization. In these circumstances, the strict consensus tree is the preferred topology because it is both of minimal length and all its resolved nodes are supported by data; that is, it is the strictly supported cladogram (Nixon and Carpenter 1996b).

7.3 COMBINABLE COMPONENTS OR SEMI-STRICT CONSENSUS

Bremer's (1990) discussion of Nelson's (1979) paper drew attention to the

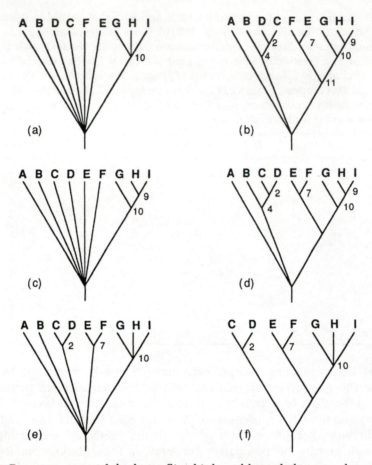

Fig. 7.2 Consensus trees of the butterflies, birds and bats cladograms shown in Fig. 7.1. (a) Strict consensus tree. (b) Majority-rule consensus tree/Median consensus tree. (c) Combinable components or semi-strict consensus tree. (d) Nelson consensus tree. (e) Adams consensus tree. (f) Greatest agreement subtree. Consensus trees were calculated using COMPONENT 2.0 for Windows®. Numbers refer to components in Table 7.1.

distinction between strict consensus and combinable components consensus and showed that components that were not replicated in all fundamental cladograms, but were nevertheless non-conflicting, could also be combined, rather than collapsed into a polytomy, as is done in strict consensus. Such non-replicated but non-conflicting components can occur when at least one of the fundamental cladograms contains a polytomy. A component that is one possible resolution of a polytomy cannot conflict with the polytomy itself and can also be included in a consensus tree. Thus, a combinable component consensus tree is formed from all of the uncontradicted components from a set of fundamental cladograms. In combinable component consensus, components need not be present in all cladograms to appear in the consensus tree. Thus, in our butterfly, bird and bat example (Fig. 7.1), combinable components consensus results in a tree (Fig. 7.2c) with slightly greater resolution than was achieved with strict consensus, in that it includes component 9 (H + I). This component is one possible resolution of the trichotomy, GHI, and therefore cannot be in disagreement with the bat cladogram, in which taxa G, H and I are unresolved. When all the fundamental cladograms are fully resolved, with no spurious resolution due to ambiguous optimization, then the combinable components consensus tree and the strict consensus tree will be identical.

7.4 MAJORITY-RULE AND MEDIAN CONSENSUS TREES

When many trees are being compared, a majority-rule consensus tree may be preferable to strict consensus (Swofford 1991). Instead of including only those groups that occur in the entire set of fundamental cladograms, it is possible to retain a pre-specified number of those cladograms in the majority-rule consensus tree. Typically, majority-rule trees are specified to retain those components that occur in more than 50% of the cladograms (Margush and McMorris 1981). Thus, when set at 50%, the groups retained must appear in *more* than half of the fundamental cladograms, because different groups that occur in exactly 50% of the trees may conflict with each other. In our butterfly, bird and bat example (Fig. 7.1), the four components shared by the butterflies and birds (Table 7.2, components 4, 7, 9, 11), together with the universal component 10, appear in the majority-rule consensus tree (with percentages of 66% and 100% respectively) (Fig. 7.2b).

The median consensus procedure (Barthélemy and Monjardet 1981, Barthélemy and McMorris 1986) is closely related to majority-rule consensus method (Swofford 1991, Page 1993b). This method uses a tree comparison metric that measures the degree of disagreement between any pair of cladograms. Thus, if the distance between a pair of cladograms, T_i and T_j, is $d(T_i, T_j)$, then the total distance of any cladogram T to k rival cladograms is given by:

$$d_T = \sum_{i=1}^{k} d(T, T_i)$$

A cladogram, T_m, is a median tree if its total distance, d_m, to the rival trees is less than that for any other cladogram. If the symmetric difference distance (Robinson and Foulds 1981, Hendy *et al.* 1984) is used as the tree comparison metric, then the 50% majority-rule consensus tree is a median tree

Consensus trees

Table 7.2 Component compatibility matrix. Numbers refer to the components in Table 7.1. Full stops (.) designate incompatible components, i.e. those that show conflict; 1s represent agreement between pairwise comparisons. The asterisks indicate the two components (9 and 10) that are uncontradicted on all three cladograms.

Compatibility matrix											
1	–										
2	.	–									
3	.	1	–								
4	1	1	.	–							
5	1	1	1	1	–						
6	.	1	1	.	.	–					
7	1	1	1	1	1	.	–				
8	.	1	1	.	.	1	1	–			
9	1	1	1	1	1	1	1	1	–		
10	1	1	1	1	1	1	1	1	1	–	
11	1	1	1	1	1	.	1	.	1	1	–
12	1	1	.	1	.	1	1	1	1	1	1	–
	1	2	3	4	5	6	7	8	9	10	11	12
	–	–	–	–	–	–	–	–	–	–	–	–
									*	*		

(Barthélemy and McMorris 1986, Swofford 1991) (Fig. 7.2b). When there is an odd number of rival cladograms (k) or if there are no groups that appear in exactly 50% of the rivals, the majority-rule tree is the only median tree. However, when k is even, any tree representing a combination of the majority-rule consensus tree with one or more combinable groups that occur in exactly half of the rival cladograms is also a median tree (Swofford 1991). Of the methods described here, median consensus trees are the least frequently encountered.

7.5 NELSON CONSENSUS

There has been some controversy as to what exactly is Nelson consensus. Some have mistaken it for strict consensus (see Page 1989), while Nelson's (1979) own presentation seems most like combinable components consensus (Nixon and Carpenter 1996b). Here, we adopt the notion of Nelson consensus proposed by Page (1989), who likened Nelson's discussion of the relationships among components to clique analysis. Page extended Nelson's concept into a method for combining components that do not include contradictory replicated sets, i.e., a formal clique analysis of cladogram components. Thus, components may appear in the Nelson consensus tree that can be contradicted in some of the original cladograms. This version of Nelson consensus is available as part of the COMPONENT for Windows® computer package (Page 1993*b*). By way of example, we can examine the result obtained when determining the largest cliques for our butterfly, bird and bat example (Fig. 7.1). If we consider taxon B, it occurs as the sister taxon to taxa C and D in the butterfly clade, B(CD), as the sister to taxon D in the bird clade, C(BD), and as the sister to taxon F in the bats clade, E(FB). As the sister pair CD occurs in both the butterflies and bats (i.e., in 66% of the cladograms), and the group BCD occurs in the butterflies and the birds (also in 66% of the cladograms), then the largest clique for all three cladograms is B(CD). For our butterfly, bird and bat example (Fig. 7.1), the Nelson consensus tree (Fig. 7.2d) is identical to the majority-rule tree, with components 2, 4, 7, 9, 10 and 11 comprising the largest clique for all three cladograms (Table 7.2).

7.6 ADAMS CONSENSUS

Adams consensus trees (Adams 1972) are derived by relocating those taxa that occur in conflicting positions on different fundamental cladograms to the nearest node they have in common. Adams trees therefore contain all the intersecting sets of taxa (nestings) common to all the fundamental cladograms in any given set of cladograms. Thus, given two sets of branches, A and B, and a cladogram *T*, set A nests inside set B if A is a subset of B and clades in A have a more derived common node in *T* than does set B. For example, given the cladogram A(B(C, D)), the sets BC, BD and CD all nest inside ABCD, but only CD nests inside BCD (Page 1993*b*). Adams trees are particularly useful for summarizing similarities in topology in the fundamental cladograms when they contain one or more taxa that show very different positions. For example, in our butterfly, bird and bat example (Fig. 7.1), all three cladograms share the nestings CD (component 2), EF (component 7) and GHI (component 10). However, taxa A and B have three completely different placings and thus are placed on the Adams consensus tree at the lowest common node, that is, the unresolved basal node (Fig. 7.2e). Adams consensus trees must be used with care because components can appear in them that do not occur in any of the fundamental cladograms.

7.7 AGREEMENT SUBTREES OR COMMON PRUNED TREES

In all the consensus methods described above, the consensus tree contains the same number of taxa as do the fundamental cladograms. A rather different method of comparing cladograms is an agreement subtree, which shows only the clades *and* taxa that are common to two or more fundamental cladograms. In this method, the greatest agreement subtree (GAS) is obtained by pruning one or more branches from each fundamental cladogram until a set of identical topologies is obtained. Finden and Gorden (1985) referred to these as 'common pruned trees'. The GAS is the subtree that results from pruning the least number of branches from the fundamental cladograms. Given two cladograms, T_1 and T_2, Page (1993*b*) defined the distance, $d_{GAS}(T_1, T_2)$, as the number of branches removed to obtain the greatest agreement subtree. By using a recursive algorithm (e.g. Kubicka *et al*. 1995), a branch of one cladogram is selected and compared to the other cladograms. The largest supported subtree in the other cladograms of comparison that contains the selected branch is maintained. Each branch is selected in turn and compared to the other cladograms. From amongst the full range of subtrees obtained the largest agreement subtree is selected.

This method is most useful when one or two taxa are incongruent amongst the profile of fundamental cladograms. Rosen (1979) used a common pruned subtree to indicate common components in two seemingly different area cladograms for central American fishes. In our butterfly, bird and bat example (Fig. 7.1), the largest agreement subtree is that which has had taxa A and B pruned out to leave a common topology of six taxa (Fig. 7.2f). An alternative implementation might permit the inclusion of uncontradicted components, in which case, the trichotomy in Fig. 7.2e could be resolved as G(HI).

7.8 CONCLUSIONS

It should be recognized that in most studies, it is the fundamental (data-derived) cladograms that provide the most direct and reliable evidence of relationships among the taxa under study. The only exception to this rule is when the strict consensus tree is the same length as the fundamental cladogram and is thus the strictly supported cladogram for that data set. Then it represents our best estimate of relationships among the taxa. In all other instances, when the consensus trees are longer than the fundamental cladograms from which they are derived, they are worse estimates of taxon relationships. However, of the plethora of methods and statistics for comparing fundamental cladograms, consensus trees are the most useful for examining the results of cladistic analyses that yield more than one minimum length cladogram. Consensus trees provide summary information about among-cladogram character conflict by providing an upper bound for the length of characters among equally most parsimonious cladograms (Nixon and Carpenter 1996*b*). Strict consensus trees include only those groups (components) for which there is unambiguous support among fundamental cladograms. All other methods include some degree of ambiguous support.

7.9 CHAPTER SUMMARY

1. When a cladistic analysis produces more than a single most parsimonious cladogram, consensus analysis provides a convenient means for summarizing agreement and disagreement among these cladograms.

2. Cladograms derived from a data set directly are called fundamental because they summarize its hierarchical information. In contrast, consensus trees are derivative, as they are constructed from and represent a set, or sets, of cladograms.

3. If the goal of consensus analysis is to summarize agreement in grouping among the fundamental cladograms, then only strict consensus fulfils that goal. All other methods may yield trees that are not fully supported by data or are supported only ambiguously. These have been called compromise trees (Nixon and Carpenter 1996*b*).

4. A strict consensus tree includes only those groups (components) that occur in *all* the fundamental cladograms. If the strict consensus tree is the same length as its fundamental cladograms, then the extra resolution present in those cladograms is spurious and due to ambiguous character optimization. In these circumstances, the strict consensus tree represents the preferred solution because it is both of minimal length and has all its resolved nodes supported by data, that is, it is the strictly supported cladogram.

5. A combinable components (or semi-strict) consensus tree is formed from all the uncontradicted components from a set of fundamental cladograms. In other words, it includes all of the components found in the strict consensus tree but in addition includes those components that are uncontradicted by less resolved components within the set of fundamental cladograms. When all the fundamental cladograms are fully resolved with no spurious resolution due to ambiguous optimization, then the strict and combinable components consensus trees will be identical.

6. A majority-rule consensus tree includes only those components that occur in more than 50% of the fundamental cladograms. When there are only two fundamental cladograms, the majority-rule and strict consensus trees are identical.

7. The median consensus method is closely related to majority-rule consensus but uses a tree comparison metric to measure the degree of disagreement between any pair of cladograms.

8. A Nelson consensus tree consists of the clique of mutually compatible components that are most replicated in the fundamental cladograms. When there are only two fundamental cladograms, the Nelson and combinable components consensus trees are identical.

9. An Adams consensus tree contains all the intersecting sets of taxa common to the fundamental cladograms. Taxa in conflicting positions are relocated to the most inclusive node they have in common. Consequently, components can appear in an Adams consensus tree that do not occur in any of the fundamental cladograms. Adams consensus is most useful for identifying 'rogue' taxa that can occur in many disparate positions in the fundamental cladograms and thus cause the strict consensus tree to be highly unresolved.

10. A greatest agreement subtree or common pruned tree differs from all other consensus trees by including only the components *and* taxa held in common by the fundamental cladograms. It is obtained by pruning one or more branches from each fundamental cladogram until a common topology is obtained.

8.

Simultaneous and partitioned analysis

8.1 INTRODUCTION

An area of cladistic methodology that has attracted much attention recently concerns procedures for treating data derived from different sources. Most systematists acknowledge that there are different kinds of data, e.g. morphological, molecular, embryonic or larval, behavioural, etc. The debate concerns the methods by which we analyse these data and combine them to reveal a common phylogenetic history. Some authors (e.g. Kluge 1989) argue that all data should be analysed in a single matrix (Fig. 8.1a). This has been called the total evidence or character congruence approach because the final clado-gram(s) results purely from interaction among all available characters. A subtle but important change in terminology, which we adopt, has recently been introduced by Nixon and Carpenter (1996a), who argued that this approach is best called simultaneous analysis because all systematists ideally use all (total) evidence, irrespective of the way in which they then deal with that evidence. Other authors (e.g. Miyamoto and Fitch 1995) prefer to analyse data separately and then use consensus methods to combine the resulting cladograms (Fig. 8.1b). This is called partitioned analysis or the taxonomic congruence approach because the final cladogram(s) is the result of adding together taxon cladograms, each derived from analysis of separate data sets for the same taxa. Although there are advocates of both approaches, some systematists have been more concerned with prescribing conditions under which one or the other method may be most appropriate. That is, in any particular circumstance, should we combine data (simultaneous analysis) or keep them separate (partitioned analysis)? In this chapter, we will outline the main theoretical and methodological basis of both approaches, the claimed advantages of each, as well as discussing possible conditions under which it may be better to use either simultaneous or partitioned analysis.

It is worth emphasizing that vicariance biogeography relies on a partitioned analysis approach inasmuch as cladograms of different organisms inhabiting similar areas are 'added' together using consensus techniques. However, in this chapter, we are concerned solely with analysis of different kinds of data relevant to the phylogenetic history of a particular group of organisms.

By way of introduction, here are some examples of studies that have used both approaches to the same taxonomic problem.

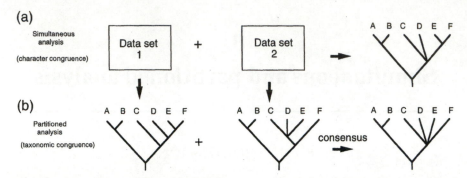

Fig. 8.1 (a) In simultaneous analysis, separate data sets are combined into a single matrix before being analysed. (b) In partitioned analysis, each data set is analysed separately (vertical arrows), yielding sets of intermediate cladograms that are then 'added' together using a consensus method. In both approaches, if more than one most parsimonious cladogram is obtained from the analysis of a data set, then consensus may be used to summarize the results.

Example 1: Milkweed butterflies

Vane-Wright *et al.* (1992) coded morphological and chemical characters expressed by ten species of African Milkweed butterflies. The chemical characters were coded as the presence/absence of volatile pheromone components in the male abdominal hairpencils. Separate analyses of the morphological and chemical data each found three equally most parsimonious cladograms. The strict consensus of each analysis is shown in Fig. 8.2. Combining these two consensus trees to give an overall strict consensus tree resulted in the solution shown at top right, which identified three monophyletic groups (circled). When the morphological and chemical characters were combined into a single data set and subjected to simultaneous analysis (bottom right), a single cladogram was found that identified eight monophyletic groups (circled). Therefore, in this study, simultaneous analysis led to far greater resolution than did the partitioned evidence approach and, incidentally, this unique solution was identical to one of the three alternative morphological cladograms. Why this was so is explained later in this chapter.

Example 2: Echinoid phylogeny

This example (Fig. 8.3) is taken from a larger study that sought to establish relationships among Recent and fossil sea urchins (Littlewood and Smith 1995). This extract deals only with those Recent taxa for which both morphological and molecular information are available. Three data sets were used: morphology, LSU (large subunit) rRNA, and SSU (small subunit) rRNA. As in example 1, simultaneous analysis led to far greater resolution than did partitioned analysis. In the cladograms derived from the separate analyses (Fig. 8.3, top), note the very different position of *Arbacia*. This was later identified by Littlewood and Smith (1995) as a 'rogue taxon'.

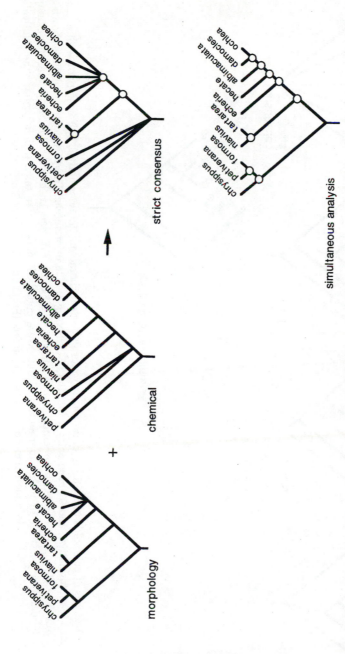

Fig. 8.2 Analysis of seven milkweed butterflies of the genus *Amauris* and three outgroup taxa (*Danaus chrysippus*, *Tirumala formosa* and *T. petiverana*). Two data sets were available, comprising 32 morphological characters (29 informative) and 68 chemical characters (63 informative). The partitioned analysis is shown along the top row with the strict consensus trees from the two separate analyses and the combined strict consensus tree. The result of simultaneous analysis is shown in the bottom row. Informative nodes are circled. (From Vane-Wright *et al.* 1992.)

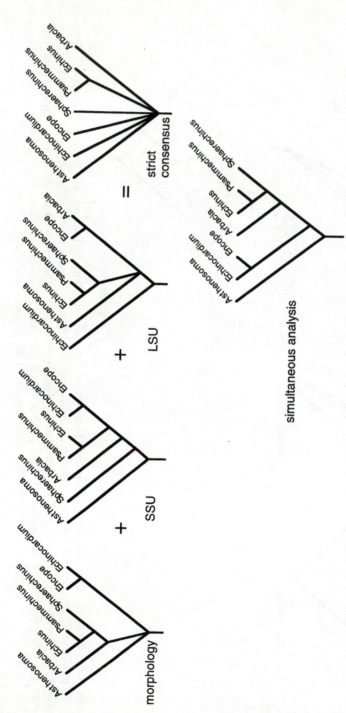

Fig. 8.3 Analysis of echinoids using three data sets: 163 morphological characters (50 informative), 278 base pairs within the SSU rRNA gene (34 informative), and 91 base pairs within the LSU rRNA gene (28 informative). The partitioned analysis is shown along the top row. The cladogram derived from simultaneous analysis of all three data sets is shown below. Note that this example is extracted from a larger study in which morphological data were coded for many more taxa, which explains the relatively few informative morphological characters. (From Littlewood and Smith 1995.)

Example 3: Deer mice and grasshopper mice

This example (Fig. 8.4) is taken from a study by Sullivan (1996), who compared the results of partitioned and simultaneous analysis of two mitochondrial gene sequences, cytochrome b and 12S rRNA. Sullivan was interested in the performance of these data sets as judged against a particular phylogeny (taken as true). The results from the separate analysis of each gene sequence were each reasonably well resolved, with strong bootstrap support for all nodes (but see Chapter 6 for a discussion of the flaws of this statistic). However, despite this, when the two cladograms were combined into a strict consensus tree, many components were lost. In contrast, simultaneous analysis yielded a single cladogram that was congruent with both the cytochrome b result and the 'true phylogeny'.

8.2 THEORETICAL ISSUES

Both approaches to data analysis are based on different premises. Simultaneous analysis accepts that the goal of cladistics is to maximize explanatory power; that is, it aims to explain the distribution of all available characters in the most parsimonious fashion (maximum informativeness). The issue of whether this yields the true (but usually unknowable) phylogeny is a separate issue. We may also look at this in a slightly different way. Any phylogenetic analysis attempts to maximize homology. Our method of testing hypotheses of homology is through character congruence (see Chapter 2), from which it follows that the more characters are included as potential tests of homology, the more severe that test becomes and the more confident we may become in accepting the truth of the conclusion.

Partitioned analysis starts from the premise that there are genuinely independent types of data (both in kind and ability to yield a phylogenetic signal), which need to be analysed separately. The most extreme examples may be molecular and morphological data where, in some cases gene cladograms do not match species cladograms. One rationale for partitioned analysis states that the results of separate analyses may provide tests for one another. That is, by looking for repeating patterns, we may observe confirmation or rejection of an original hypothesis (Grande 1994). The underlying assumption is that different data sets are truly independent indicators of relationship.

Most of the debate between the advocates of simultaneous and partitioned analysis has concerned justification for different methodologies and the style of argument has been that of claim and counterclaim. Rather than treat each view separately, we will try to avoid repetition and take the alleged principal advantages of both and discuss both sides of the issues involved. As the advocates of partitioned analysis have raised more issues, we begin with this approach.

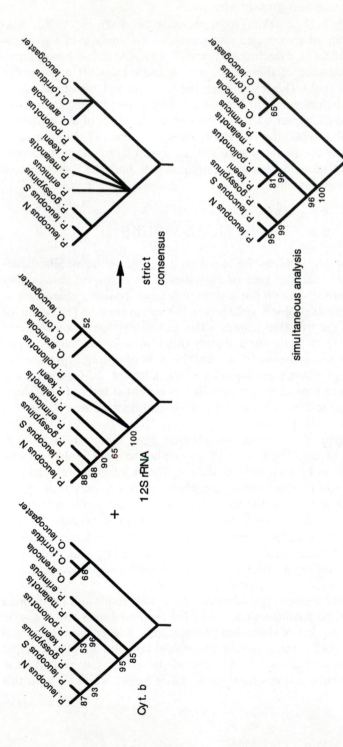

Fig. 8.4 Analysis of two mitochondrial genes of seven species of deer mice (*Peromyscus*) and three species of grasshopper mice (*Onychomus*). The cytochrome b (Cyt. b) sequence contained 61 informative characters and the 12S rRNA had 48 informative characters. The partitioned analysis is shown along the top row. The cladogram derived from simultaneous analysis is shown below. The figures are the percentage of bootstrap replicates in which the groups were recovered. (From Sullivan 1996.)

8.3 PARTITIONED ANALYSIS (TAXONOMIC CONGRUENCE)

8.3.1 Independence of data sets

The crux of arguments in favour of partitioned analysis rests on the belief that there are genuinely different classes of data, which may reflect different evolutionary processes of character change and which behave differently with respect to phylogeny reconstruction. This idea has its roots in the early literature on phenetics (Sokal and Sneath 1963). The most frequently recognized independent data sets are molecular sequences that lead to gene cladograms, and morphological data that lead to species cladograms. These may or may not agree. Some authors (e.g. Miyamoto and Fitch 1995) have suggested that in sexually reproducing organisms, the maternally inherited mitochondrial genes constitute data independent of the biparentally inherited nuclear genes. Further, protein-coding and non-coding genes may be considered as distinct and, perhaps, transcribed and non-transcribed parts of the genome may also be independent. Within the morphological realm, there may be a case for considering larval characters quite independently of adult characters, particularly in organisms that exhibit drastic metamorphosis.

Bull *et al.* (1993) defended data partitioning using what they termed 'process partitions', by which they meant subsets of characters that are evolving according to different sets of rules. Different kinds of nucleic acids as mentioned above would fall into different process partitions. Miyamoto and Fitch (1995) went further and identified five criteria of process partitions:

- The genes are not genetically linked.

- The gene products do not interact.

- The genes do not specify the same function.

- The gene products do not interact in the same physiological pathway.

- The gene products do not regulate the expression of a gene in another process partition.

Given the complexity and integral nature of gene expression, coupled with our relative ignorance of gene interaction, it may be difficult to partition data into sets according to these strict process criteria. But the intention of partitioning data is to try to establish whether we have independent means or independent data to estimate phylogenetic relationships. The observation that we may retrieve strongly supported but non-congruent topologies from different data sets is evidence that one or more is positively misleading when analysed using parsimony. The causal explanation is usually the suggestion that different evolutionary processes are acting upon different 'process partitions' with the expectation that different evolutionary signals will result. This remains to be demonstrated empirically but, if correct, then the corollary is that we would analyse separate partitions using different models of evolutionary change. Then our analyses would shift from a parsimony approach towards a maximum likelihood approach, where analyses are driven by one or more of a myriad of process models, each of which in itself requires justification. If the justification for partitioning data is to account for different evolutionary processes, then we may have difficulty deciding how fine the partitions must be. Once again the problem is most transparent in molecular studies where there are noticeable differences in evolutionary rates between and even within genes (e.g. stem and loop regions).

Bull *et al.* (1993) carried out simulation experiments in which they began with a given phylogeny. The data were then divided into two partitions, one of which was allowed to evolve quickly and the other slowly. From the results of these simulations, they claimed that combining the resultant two data sets in a simultaneous analysis was less successful in recovering the given phylogeny than was analysing the two data sets separately. However, if the data sets are differentially weighted, as is often done with molecular data, then the problem evaporates and simultaneous analysis produces the more accurate result (Chippendale and Wiens 1994). In a sense, these simulation experiments are misleading because, in reality, the 'given phylogeny' is unknown and the conditions imposed on the experiments are too simplistic. Empirical examples are much more informative. Sullivan (1996) presented such a study, which showed that among-site rate variation was more or less uniformly distributed across the sequences and that partitioning the data was neither feasible nor appropriate. His study (Fig. 8.4) attempted to discover how well sequences of cytochrome b and 12S rRNA recovered a phylogeny of deer mice (*Peromyscus*) and grasshopper mice (*Onychomys*), which was accepted as the 'true phylogeny' based on several other kinds of evidence. Separate analysis of the genes gave conflicting results but each was strongly supported (at least as measured by bootstrap values). Three separate tests suggested strong heterogeneity between the data sets (i.e. it was not due to sampling error), which, according to Bull *et al.* (1993), should mean that these data sets represent different process partitions. Therefore, these data should have been kept separate and accepted as alternative but equally valid gene histories (even though both genes are mitochondrial and are only maternally inherited). Simultaneous analysis of the two genes gave the cytochrome b topology, which was the same as the 'true topology' and with even higher bootstrap values. The conclusion was that the 12S rRNA cladogram, even though strongly conflicting with the 'true phylogeny', contained a common hidden phylogenetic signal and that combining the data in a simultaneous analysis allowed that signal to 'show through'. Sullivan's (1996) results suggested that the 'phylogenetic signal can in fact be additive when data from genes with different evolutionary histories are analysed under a homogeneous reconstruction model (parsimony with equal weights)'.

Bull *et al.* (1993) were less extreme than Miyamoto and Fitch (1995) in their process partitioning. Although they recognized the possibility of different processes, they were more concerned with demonstrable heterogeneity between data sets and the conditions under which they should be combined or kept separate. We will return to this question later in the chapter.

8.3.2 Cladogram support

Another claimed advantage of partitioned analysis is that different data sets may vary in their ability to recover a phylogenetic signal and that by keeping data sets apart this may be recognized. The idea behind this claim is that different data sets may include very different levels of homoplasy. For example, data set 1 may yield one optimal cladogram plus many suboptimal cladograms that are but one step longer. Data set 2 may yield one optimal cladogram that is many steps shorter than the nearest suboptimal solution. Data set 2 may be said to contain a stronger phylogenetic signal and this recognition may allow us to select the second cladogram in preference to the first. Our choice may also be strongly influenced if the optimal cladogram from data set 2 was identical with one of the near suboptimal cladograms from data set 1. Simultaneous analysis may or may not recover the cladogram with less homoplasy but, in any event, it may not allow us to identify whether data set 1 or data set 2 is giving the stronger signal. However, if different strengths of signal are suspected, it is possible to cater for this in a simultaneous analysis by adding or deleting characters and noting which sets of characters are congruent with the conclusion.

There is another side to this argument that supports the simultaneous analysis approach. If different data sets do yield cladograms with very different levels of support then, unless the topology of both is identical, the consensus resulting from partitioned analysis will lose any evidence of support heterogeneity. This is because the results of analysing separate data sets are each regarded as having equal weight before the cladograms are added together to give the final result. Of course, it is possible to allow for this in a simultaneous analysis by applying different weights to different character partitions that make up the total matrix. However, this does not obviate the necessity to justify such a practice explicitly.

8.3.3 Different sized data sets

Advocates of partitioned analysis argue that cladograms produced from analyses of separate data sets have an equal contribution to make to the final result. One claimed advantage of this is to avoid the potential effects of very unequally sized data matrices. It is common for molecular data sets to be an order of magnitude larger than morphological data sets with the implied danger that the molecular signal, if different, will swamp the morphological signal. Theoretically this may be possible but in practice it rarely happens. In Example 2 (Fig. 8.3), there are many more potential characters in both the LSU and SSU rRNA sequences (2200 base pairs) than in the morphological data set (163 characters). However, the numbers of informative characters that may contribute to topological variants are comparable between data sets (121 molecular and 136 morphological). This situation is by no means uncommon. Exceptions may be those sequences that show strong G-C bias (see, for example, Hedges and Maxson 1996, who favoured a partitioned analysis approach when analysing amniote relationships; also the section on weighting in Chapter 5). It remains to be demonstrated empirically that data sets will swamp one another as a rule rather than exceptionally.

For a simultaneous analysis, the only way in which one portion of the combined matrix will swamp out a signal from another portion of the matrix is if the swamping signal is sufficiently strong. With molecular data, this is rarely the case, and even then, the optimal cladogram is usually followed by suboptimal solutions favouring the morphological data that are one or two steps longer. Even accepting the fact that there may be many more informative characters in a molecular data set, if there is a stronger signal from morphological data, it will generally show through.

8.4 SIMULTANEOUS ANALYSIS

8.4.1 Resolution

One claim of the simultaneous analysis approach is that the resulting clado-gram(s) is nearly always more highly resolved than is a consensus of separate cladograms. This was so in all the examples given at the beginning of this chapter. Expressed another way, simultaneous analysis leads to greater ex-planatory power and this is the primary justification for the method. There have been no extensive counterclaims made against this point but it has been argued that even though the consensus cladogram may be less resolved, it does provide a conservative estimate of phylogeny (Swofford 1991). We interpret this to mean that by taking the conservative route, there is less danger of selecting a 'wrong' answer—which indeed may be true. Notice, however, that the contrast made here is that between character congruence and phylogeny reconstruction. The claimed advantage of partitioned data and consensus is that a conservative phylogeny (a statement of historical events) is preferable to character congruence (a statement of the most parsimonious distribution of characters).

But there are problems. The purpose of a phylogenetic tree, as opposed to a cladogram, is to plot character changes (evolution). As Nixon and Carpenter (1996a) pointed out, the conflicting cladograms that produced the polytomies in the consensus tree contain conflicting character optimizations.

Consequently, the conservativeness that is claimed for consensus trees actually means ambiguity in ideas of character evolution, which must then be resolved some other way (choosing one of the cladograms). Using a consensus tree as a conservative statement of phylogeny requires that we ignore both the taxa involved in the ambiguity and any inferences of character evolution of those characters that are optimized differently on alternative cladograms. In a consensus tree, the only statements that we can make about evolution will be limited to unambiguous character optimizations. The more highly resolved the cladogram, the more statements can be made about character evolution when the cladogram is interpreted as a phylogenetic tree. To retain all taxa in a phylogenetic analysis resulting in a consensus requires choosing one tree. Clearly, the choice is easier if there are fewer alternatives.

There could be some advantage in maintaining separate data sets in order to recognize the patterns of resolution of different kinds of data. In the *Amauris* example, analysis of both morphological and chemical data each produced three optimal cladograms. However, while for the morphological data the ambiguity involved *Amauris hecate, A. albimaculata, A. damocles and A. ochlea*, for the chemical data it involved *Danaus chrysippus, Tirumala formosa* and the genus *Amauris* as a unit. Such information would be concealed by simultaneous analysis, yet this might be of interest when comparing separate analyses of different groups of organisms using similar types of data. It may be recognized that molecular data derived from a particular gene consistently gave poor resolution amongst taxa inferred to be of particular age.

8.4.2 Arbitrary consensus

Another claimed advantage of simultaneous analysis is that it avoids arbitrariness in the consensus methods used (Kluge and Wolf 1993). However, both partitioned and simultaneous analysis use consensus trees to summarize information when more than one cladogram results from the analysis of a given data set, and to this extent there is no difference. However, partitioned analysis also uses consensus at a second, later stage (Fig. 8.1) to summarize the information common to the separate data sets, and it is possible that in order to maximize the effectiveness of this consensus, the final result is a hybrid of two or more consensus methods. There are many consensus methods (see Chapter 7), each designed to summarize information in a different way. Some (e.g. Adams consensus trees) may contain topological variants not part of the original set, while others (e.g. majority-rule consensus trees) ignore some topological variants that may be common to all separate analyses. Clearly, to combine an Adams consensus tree for data set 1 with a strict consensus tree for data set 2 and a majority-rule consensus tree for data set 3 is to compound *ad hoc* assumptions and potentially eliminate some cladograms. However, as long as the same consensus method is used through-out and justified, then the choice of consensus method is not relevant to the debate between partitioned and simultaneous analysis.

8.5 CONDITIONAL DATA COMBINATION

As mentioned at the beginning of this chapter, some authors acknowledge a utility in both approaches to data analysis and have recognized there may be occasions when one or the other is preferable. This has been called con-ditional data combination (Huelsenbeck *et al.* 1996). The 'conditional' is centred on measures of heterogeneity between the data partitions. As Bull *et al.* (1993) explained, when heterogeneity between data sets yields significantly different phylogenetic estimates that are too great to be explained by sam-pling error of either taxa or characters, then the analyses should be kept separate and simultaneous analysis not undertaken. What might provide a test? Several have been proposed. De Queiroz (1993) suggested using boot-strap values as a criterion. If, when comparing bootstrap values of conflicting clades, these values are both high, then the data should not be combined. However, there are problems with the bootstrap (see Chapter 6) and, as the study by Sullivan (1996) described above showed, it does not necessarily follow that conflicting cladograms, each of which has high bootstrap values assigned to the internal branches, give poor results when combined in a simultaneous analysis.

Other, more complex statistics have been devised to test whether incongru-ence between data sets is greater than expected by chance alone (e.g. Farris *et al.* 1994, Huelsenbeck and Bull 1996). Here, we choose Templeton's non-parametric test, as applied by Larson (1994), as an example. In this test, the most parsimonious cladogram (or cladograms) is derived for each data set. Next, the fit of characters from one data set is compared on the two alternative topologies (or on those most closely comparable if each data set produces more than one optimal cladogram). If the characters show a significantly better fit to their 'own' topology than to the 'rival' topology, then the differences between the two data sets cannot be explained as the chance result of sampling error. This test is worked through using the Milkweed butterfly example (Table 8.1). In this example, the conclusion is that the data sets can be combined because the differences between them in one of the data sets could be due to chance alone and are not the result of a significant difference in phylogenetic signal.

This and other such tests may be statistically elegant but remain relatively crude, and there are no hard and fast rules for deciding what strategy to take. Unfortunately, it is not just character sampling that may be misleading, but also taxon sampling. In the example 2 used here (Fig. 8.3), application of Templeton's test recorded differences between the morphological cladogram and SSU rRNA cladogram that were judged as insignificant, but the differences between the LSU rRNA cladogram and morphology were judged as significant, suggesting that they should not be combined. However, Littlewood and Smith (1995) recognized that this significance was due almost entirely to the different positions of *Arbacia* on the two cladograms and regarded *Arbacia* as a rogue taxon. They considered that because the heterogeneity between the data sets was caused by a single taxon, then simultaneous analysis was the preferred option.

Table 8.1 Templeton's test of data heterogeneity applied to the *Amauris* butterfly data. For each of the three fundamental cladograms that form the morphology consensus and the chemical consensus, the topology is chosen from one set that most closely resembles one from the other set. Morphological characters are then optimized on to both cladograms. For each character the difference in performance on its own cladogram and the rival cladogram is noted together with the number of extra (+) or fewer (−) steps. The total number of characters is noted keeping extra and fewer steps separate. The differences are then ranked (mid-point scores given when two or more characters share the same number of differences). The sums of the positive and negative ranks are calculated separately and the smaller of the two figures taken as the test statistic. The probability value is read from a critical value table of the Wilcoxon Rank Sum. If the probability is < 0.05 (the critical value), then the difference between the performance of the characters on to its own cladogram and the rival cladogram is significant. The reciprocal operation is also performed for the biochemical characters. In this case, it is concluded that the chemical data are not optimized significantly better on their own cladogram than on the morphological cladogram. However, the morphological data are optimized significantly better to their own cladogram than to the chemical cladogram. This means that the chemical signal is weak and the data may be combined.

| Morphology character | Number of times character changes on | | Difference | | Rank | |
	morphology topology	biochemical topology	+	−	+	−
5	1	2	1	0	5.5	0
6	1	2	1	0	5.5	0
7	1	2	1	0	5.5	0
8	1	2	1	0	5.5	0
9	1	2	1	0	5.5	0
12	1	2	1	0	5.5	0
13	1	2	1	0	5.5	0
26	1	2	1	0	5.5	0
28	1	2	1	0	5.5	0
29	1	2	1	0	5.5	0
Total = 10			10	0		
Ranked sum					55	0
Test statistic = 0; probability value < 0.05, significant						

Table 8.1 (*Continued*)

Biochemical character	Number of times character changes on		Difference		Rank	
	biochemical topology	morphology topology	+	–	+	–
37	3	4	1		6	
39	1	2	1		6	
62	1	2	1		6	
64	1	2	1		6	
65	1	2	1		6	
70	2	3	1		6	
71	1	2	1		6	
72	3	2		1		6
73	2	1		1		6
78	3	2		1		6
89	1	2	1		6	
Total = 11			8	3		
Ranked sum					48	18

Test statistic = 18; probability value > 0.05, not significant

If, as a result of a heterogeneity test, simultaneous analysis is 'recommended', then the data are combined. If, however, it is not 'recommended' then we must find the source of the heterogeneity. And if the heterogeneity is due to sampling then, theoretically, this may be corrected. If it is thought that it is due to different evolutionary processes then we must choose between data sets—but this is a decision taken beyond cladistic analysis.

8.6 OPERATIONAL DIFFICULTIES

In attempts to include as much data as possible in any phylogenetic analysis, we may encounter problems with both taxon and character sampling. Such difficulties will affect both partitioned and simultaneous analysis. Different taxa may have been sampled for different characters. For partitioned analysis, the problem of combining cladograms with different but overlapping sets of terminal taxa is analogous to problems encountered in cladistic biogeography, and the resolution of such dilemmas resides in the methods of component analysis. Of more immediate concern are the problems encountered in trying to combine into a common matrix different data sets that may not include the same taxon sample or representation of character completeness (e.g. fossils with Recent taxa).

Some workers reduce the taxon sampling within each data set such that the same terminal taxa are represented. But this may be too restrictive (it would certainly discriminate against including fossils in morphology and molecular studies) and could result in very poor taxon sampling for the group under study. Another strategy would be to use 'hybrid' taxa. These may be of several kinds. Perhaps the simplest situation would be to use morphological data from one species and add molecular data from another. For a very speciose genus, this may be unsatisfactory, but could be checked later when more data become available. Often, it happens that morphological data exist for many species but molecular sequences for only one. Thus, another approach might be to carry out an initial analysis of the species using morphological data and then code the 'ancestral' conditions into a morphotype, before adding the single molecular sequence. Here the hybrid is between a real and inferential sample of characters (see Chapter 3 for the problems associated with using such artificial 'groundplans').

Until recently, it was commonplace for a sequence for a given gene to be available for only a single species within a genus and often for just a single individual within a species. However, multiple sequences are now becoming increasingly available for the most popular genes (e.g. 28S rRNA, 18S rRNA, cytb, COII) and this is giving rise to another practical sampling problem. Some species may still be represented by only a single sequence, but for others, there are several sequences known that, in addition, vary among themselves. Then, in order to avoid polymorphic codings for some taxa, variable sites could be deleted in all terminal taxa.

A rather special case of data diversity occurs when fossils are included in analyses embracing data from soft anatomy as well as molecular sequences. Here, there would be many question marks inserted against the fossil taxa. The disruptive effect of many question marks has already been considered in Chapter 4. Discussion of this topic has focused on situations in which the question marks are scattered throughout the data matrix and there have been fears that many question marks may cause convergence onto an incorrect solution (Huelsenbeck 1991*b*). However, in situations where we may wish to add fossils that are reasonably well known from skeletal data but are unknown for other types of data, all the question marks are clustered in one partition of the matrix. Recently, Wiens and Reeder (1995) carried out simulation tests on real data and showed that when missing data are concentrated, the effect on inaccuracy is relatively minor. This gives us hope that including fossils, which have the ability to break up long branch lengths, may be more beneficial than excluding them.

8.7 CONCLUSIONS

In many respects, devotees of partitioned analysis and simultaneous analysis are seeking different goals, and much of the discussion of one faction is actually tangential to the position of the other. Most arguments in favour of partitioned analysis are concerned with the recognition that evolutionary processes may be affecting different data in different ways. The goal is to recognize those partitions so that they may be analysed using different evolutionary models. The perceived acceptable risk is relaxation of the primary cladistic principle of parsimony. The strength of partitioned analysis lies in the estimation of the reliability of phylogenetic signals coming from different data sources. Partitioned analysis is concerned with evolutionary trees and patterns of character evolution, even though the tree may be poorly resolved.

Simultaneous analysis is driven by the goal of parsimony over all characters, which maximizes information content and provides the most severe test of homology. The primary aim is to establish a cladogram, which can then be converted to a phylogenetic tree from which we can infer evolutionary pattern and process. The acceptable risk is that it may disguise the relative strengths of different phylogenetic signals. Since the discovery of homology is the central tenet of this book, we recommend simultaneous analysis as the appropriate approach for cladistic analysis.

8.8 CHAPTER SUMMARY

1. Simultaneous analysis combines all available data, from whatever source, into a single data set for analysis.

2. Partitioned analysis allocates data derived from different sources to separate data sets, which then are analysed individually before combining the results using a consensus method.

3. Simultaneous analysis aims to maximize explanatory power of the data by explaining the distribution of all characters in the most parsimonious manner. By testing hypotheses of homology through character congruence, it follows that the greater the number of characters included in an analysis, the more rigorously are those hypotheses tested. This is the strength of simultaneous analysis.

4. In contrast, partitioned analysis is based upon the premise that there are genuinely different classes of data that may reflect different evolutionary processes. These data classes may thus be independent indicators of relationships and need to be analysed separately. In this way, the results of the separate analyses may provide tests of one another.

5. Classes of characters that evolve according to different rules are termed process partitions. However, given the complexity of gene expression and our meagre knowledge of evolutionary processes, it is very difficult to justify where to draw up the partitions or to decide how fine they must be.

6. Different data sets may vary in their levels of homoplasy and thus in their ability to recover a phylogenetic signal. By keeping data sets separate, this may be recognized. This is the strength of partitioned analysis.

7. It has been claimed that combining a data set with a small number of characters with one that is much larger may lead to the swamping of its phylogenetic signal. However, in practice, this has rarely been found to occur.

8. Simultaneous analysis often produces a more resolved result than does partitioned analysis. Thus, simultaneous analysis leads to greater explanatory power.

9. Conditional data combination is based upon the premise that there may be some occasions on which simultaneous analysis should be undertaken and others when the partitioned approach may be preferable. The heterogeneity in phylogenetic signal between data sets is calculated and if this is greater than can be explained by sampling error, then the data sets should be kept separate. Heterogeneity tests include the use of the bootstrap and Templeton's test.

10. If data from two or more sources are unavailable for some taxa, then combining data into a single matrix may lead to a large percentage of missing values. However, in such situations, the missing values are generally concentrated in one partition of the matrix and simulation tests have shown that their disruptive effects are relatively minor.

9.
Three-item statements analysis

9.1 INTRODUCTION

De Pinna (1996) wrote that 'the most interesting idea in mainstream theoretical systematics in recent years is the so-called three-item analysis'. This may strike cladists as a curious statement, given that most commentary relating to three-item statements analysis has been decidedly negative (Harvey 1992; Kluge 1993, 1994; Farris *et al.* 1995). He went on to point out that a major success of the cladistic approach was the recognition that ancestor–descendent relationships among taxa cannot be objectively proposed or tested, only sister-group relationships (see Chapter 1). The idea that one taxon can 'give rise' to another is acknowledged to be beyond proper scientific investigation. It is perhaps curious that cladists still treat characters in a pre-cladistic way, as if one character state can give rise to another, relating one state to another in ancestor–descendent fashion. This approach seems embedded in the standard approach to character coding with the recognition of 'transformation series' and the use of character optimization. If ancestral taxa have been discarded from scientific enquiry then why not ancestral characters? Both appear to be based on the absence of evidence and the formalism of conventional cladistic approach.

All cladists agree that cladistics is about grouping by synapomorphy and that synapomorphy is evidence of relationship. Three-item statements analysis is an alternative way to code data based on the idea that 'taxon' and 'homology' represent the same relationship (Nelson 1994). It departs from other approaches by focusing on the smallest possible unit of relationship, the three-item statement, and how these fit most parsimoniously to possible cladograms. In this sense, three-item statements analysis is an entirely different way of viewing data. According to its creators (Nelson and Platnick 1991), three-item statements analysis improves the precision of parsimony. This chapter concentrates on the implementation of three-item statements analysis and outlines some possible ways in which precision is improved.

9.2 CODING

Prior to analysis, systematic data (observations) are coded as series of binary or multistate characters reflecting judgements of primary homology (see

Chapter 2). Using a four taxon example (A–D), if taxa C and D are observed to have a feature, then they are usually coded as '1' and taxa A and B, which lack that feature, are coded as '0'. The binary character incorporates an element of 'identity' (the 1s) and an element of 'difference' (the 0s). Each binary character is assumed to be an independent homology. For multistate characters, the different states are assumed homologous among themselves and as a consequence are non-independent (i.e. the 1s, 2s, etc., are dependent). Multistate characters are often represented (or interpreted) as suites of binary characters in a cladistic analysis. For the purposes of this chapter, representation of data as binary or multistate variables will be referred to as the standard approach.

Three-item statements analysis, in contrast, does not represent systematic data as binary and multistate variables but reduces observations to their simplest expression of relationship, a three-item statement. For example, the three-item statement A(BC) implies that taxa B and C share a relationship to the exclusion of taxon A. Suites of three-item statements can be arranged into a statement × taxon matrix for analysis with a parsimony program, in an identical fashion to standard binary and multistate data.

Using the same four taxon example (A–D) as above, in which C and D

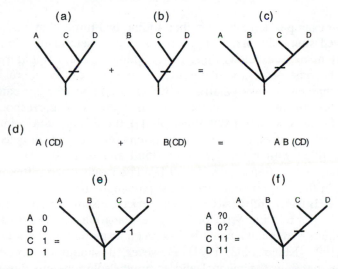

Fig. 9.1 Different analytical representations of one binary character, AB(CD), for which two three-item statements are possible: A(CD) and B(CD). (a) Diagrammatic representation of the three-item statement A(CD). (b) Diagrammatic representation of the three-item statement B(CD). (c) Diagrammatic representation of the solution to A(CD) + B(CD) = AB(CD). (d) Written representation of the statements and solution in (a–c). (e) Standard data matrix and its solution. (f) Three-item statements matrix and its solution. Note that while this solution appears to be the same as (d), it is actually a strict consensus tree of three equally most parsimonious solutions and the minimal cladogram.

possess a particular feature and A and B do not, two three-item statements are possible: A(CD) and B(CD) (Fig. 9.1a–b). Addition of these two three-item statements produces a summary cladogram (Fig. 9.1c), in which the two three-item statements combine to unite C and D: A(CD) + B(CD) = AB(CD) (Fig. 9.1d). This is identical to the summary cladogram from the binary character, AB(CD), which unites C + D on the basis of a common possession of state 1 (Fig. 9.1e).

From this simple example, we can see that for primary representation of data (the original observations), there is no difference between the three-item statements and the standard approaches. The choice facing systematists rests upon which *aspect* of the data they wish to represent from their original observations. As a starting point, it is worth remembering that cladistics, in its most general form, is concerned with hierarchical patterns, whether those patterns express relationships among characters, taxa, genes or areas (Nelson and Platnick 1981). In short, an hierarchical pattern does not imply a process but expresses degrees of relationship. Cladistics is the study of relationships.

9.3 IMPLEMENTATION

The above example demonstrates that adding two three-item statements can be achieved simply by hand. However, most data sets involve many (sometimes very many, see below) three-item statements and computerized methods then become necessary. Three-item statements can be represented in a standard character × taxon matrix. However, if we consider our earlier example (Fig. 9.1a,b), for the statement A(CD), there is no corresponding data point for taxon B, while for statement B(CD), there is no data point for taxon A. Nevertheless, current parsimony programs require that all cells in a data matrix be filled with a value of some kind and so these 'data' points are represented by question marks (Fig. 9.1f). The results then differ from those expected, in that not one but three most parsimonious cladograms are found: AB(CD), A(B(CD)) and B(A(CD)). The strict consensus tree of these three solutions is shown in Fig. 9.1f (for further explanation, see §9.3.6). This aspect of three-item statements analysis has been exploited in various critiques (Harvey 1992, Kluge 1993, 1994). However, it should be noted that the differences between the manual and computer solutions are due to current idiosyncrasies of parsimony programs, especially their treatment of question marks (see Chapter 4), and not to the form in which the data are represented (Nelson and Ladiges 1993). It should be borne in mind that implementation of a method involves issues separate from the reasons to adopt the method, although both are connected.

9.3.1 Binary characters

The three-item statements equivalent to a binary character are determined by comparing each pair of taxa that has the informative state with every taxon that lacks that state. For example, for a character expressing the relationship ABC(DE), there are three three-item statements: A(DE), B(DE) and C(DE). Only taxa DE possess the informative state and as a pair are related to A, B and C. The number of possible three-item statements is given by $(t - n)n(n - 1)/2$, where t = the total number of taxa and n = the number of taxa with the informative (apomorphic) state. For ABC(DE), $n = 2$ and $t = 5$, hence $(5 - 2)2(2 - 1)/2 = 3$ statements. For a character expressing the relationship AB(CDE) there are six three-item statements: A(CD), A(CE), A(DE), B(CD), B(CE), and B(DE). Taxa CDE possess the informative state and constitute three pairs, CD, DE, and CE with each pair related to A and B. Thus, $n = 3$ and $t = 5$, hence $(5 - 3)3(3 - 1)/2 = 6$ statements.

9.3.2 Multistate characters

Multistate characters are usually disassembled into suites of binary characters for the purposes of analysis (see Chapter 2). However, it is now recognized that such re-coding may involve redundancy. From the perspective of three-item statements analysis, a multistate character is equivalent to a suite of unique three-item statements so that no statement appears more than once in the suite (Nelson and Ladiges 1992). For example, an ordered multistate character expressing the relationship A(B(CD)) has four three-item statements, A(BC), A(BD), A(CD), and B(CD). Manually, A(B(CD)) can be reduced to its basic components: A(BCD) and B(CD). A(BCD) has three three-item statements, A(BC), A(BD) and A(CD) and B(CD) provides the fourth. How does this differ from its binary representation? The multistate character A(B(CD)) can be represented as two binary characters: A(BCD) and AB(CD). As before, A(BCD) yields three statements, A(BC), A(BD) and A(CD), while AB(CD) yields two statements, A(CD) and B(CD), giving a total of five. The two binary characters have one more statement than the single multistate character, because A(CD) occurs twice. Thus two binary characters (understood as independent) appear to have more information than the multistate character in which the states are seen as dependent. This points to a possible difference in information content between a multistate character and its binary equivalent (see below).

9.3.3 Representation of three-item statements for analysis with current parsimony programs

A three-item statement involves only three terminals, but current parsimony programs require all cells of a matrix to be filled. For example, a three-item statement pertinent to a 12-taxon problem will require nine terminals to be ignored. This requirement can be satisfied by the use of question marks in the appropriate cells. This 'value' (which is not a real value) should be interpreted as truly meaning 'non-applicable', rather than the 'either/or' or 'polymorphic' interpretation that can also be attributed to a question mark (see Chapter 4). This simple heuristic device has been persistently misunderstood by critics (e.g. Harvey 1992, Kluge 1993, 1994). However, the use of question marks may present problems in the resulting cladograms by producing over-resolution of nodes that are not supported by data. Nelson (1992) introduced the concept of 'minimal cladogram' (see below) with respect to three-item data, which is similar to the strictly supported cladograms in standard analyses (see §4.2).

9.3.4 Cladogram length and three-item statements

Table 9.1a is a standard matrix for four taxa (A–D) and 10 binary characters (1–10). Analysis of this matrix yields one cladogram (Fig. 9.2a; length = 13). The three-item statements matrix for the same data is given in Table 9.1b, analysis of which yields the same cladogram (Fig. 9.2a, length = 30). The three-item statement 'characters' can be grouped into six sets of three-item statements, with a total of 24 statements: A(CD) ($\times$7), B(CD) ($\times$3), A(BC) ($\times$4), A(BD) ($\times$4), C(AB) ($\times$3), D(AB) ($\times$3). Of these, 18 are included in the resulting cladogram, while six are excluded. The former are referred to as 'accommodated three-item statements' (ATS) and the latter as 'non-accommodated three-item statements' (NTS). Accommodated statements each fit to a node with a single step and add one step to the cladogram length.

Table 9.1 (a) Standard matrix of four taxa (A–D) coded for 10 binary characters. (b) The corresponding three-item statements matrix. Statements have been arranged into six groups of equivalent statements. Characters 1–3 and 8–10 yield two three-item statements (a and b), while characters 4–7 yield three statements (a–c) See text for further details.

(a)

	1 2 3 4 5 6 7 8 9 10
A	0 0 0 0 0 0 0 1 1 1
B	0 0 0 1 1 1 1 1 1 1
C	1 1 1 1 1 1 1 0 0 0
D	1 1 1 1 1 1 1 0 0 0

(b)

	1 2 3 4 5 6 7 a a a a a a a	1 2 3 b b b	4 5 6 7 b b b b	4 5 6 7 c c c c	8 9 10 a a a	8 9 10 b b b
A	0 0 0 0 0 0 0	? ? ?	0 0 0 0	0 0 0 0	1 1 1	1 1 1
B	? ? ? ? ? ? ?	0 0 0	1 1 1 1	1 1 1 1	1 1 1	1 1 1
C	1 1 1 1 1 1 1	1 1 1	1 1 1 1	? ? ? ?	0 0 0	? ? ?
D	1 1 1 1 1 1 1	1 1 1	? ? ? ?	1 1 1 1	? ? ?	0 0 0

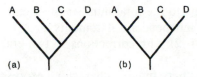

Fig. 9.2 (a) Three-item solution for the matrix in Table 9.1b under uniform weighting. (b) Alternative cladogram found using fractional weighting of the same data.

Non-accommodated statements fit the cladogram twice and add two steps to the length. Thus, the relationship between data (the suite of three-item statements) and cladogram length is given by:

$$\text{Length} = \sum \text{ATS} + 2\left(\sum \text{NTS}\right).$$

For the example in Table 9.1b, the accommodated statements are A(BC), A(BD), A(CD) and B(CD), giving a total of 18, and the non-accommodated statements are, C(AB) and D(AB), which add a further 12 steps, giving an overall cladogram length of 30, which is the length reported by parsimony programs.

The shortest cladograms accommodate the largest number of statements and are hence selected as optimal. The use of 'steps' should not be confused with the conventional understanding of that term, which is equivalent to a character transformation. A three-item statement involves only three terminals, hence it can only fit a node of a cladogram exactly (with one step) or inexactly (with two steps). The essential difference in interpretation is that the length statistic indicates 'fit' of data rather than 'character' change.

9.3.5 Uniform and fractional weighting

It was noted above that, for binary characters, the equivalent number of three-item statements is given by $(t - n)n(n - 1)/2$. However, when $n > 2$, there may be redundancy among statements. Consider, for example, the binary character expressing the relationship A(BCD), which has three statements, A(BC), A(BD) and A(CD). Combination of any pair yields the original relationship, A(BCD), which logically implies the third:

$$A(BC) + A(BD) = A(BCD)$$
$$A(BC) + A(CD) = A(BCD)$$
$$A(BD) + A(CD) = A(BCD)$$

This suggests that, in some cases, not all statements are independent. In this example, because only two of the three statements are required to reproduce the original relationship, each can be weighted to have an absolute value of two-thirds. Generally, the number of independent three-item statements is $(n - 1)(t - n)$ and their absolute value (or weight) is the ratio of independent statements to the total number of statements, or $2/n$. Tables of values for total and independent statements are provided by Nelson and Ladiges (1992, tables 1–2), with corrections to two entries in their table 1 by Williams (submitted).

Consider now another binary character expressing the relationship ABC(DE), for which there are three statements, A(DE), B(DE), and C(DE). As in the first example, the number of independent statements is 3, but their absolute value is 1. In other words, if any one of the three statements is omitted, the other two cannot be combined to yield the correct result, ABC(DE). For example, A(DE) and B(DE) sum to give only AB(DE); C is missing. All three statements are independent. Finally, consider a third binary character expressing the relationship AB(CDE), for which there are six three-item statements: A(CD), A(CE), A(DE), B(CD), B(CE), and B(DE). The number of independent statements is 4, each with an absolute value of two-thirds. Inspection of the six statements reveals two subsets of three statements, one with relationships relevant to A, the other with relationships relevant to B: A(CD), A(CE), A(DE) and B(CD), B(CE), B(DE). Further inspection reveals that within each subset, summing any two will produce a correct solution. Thus A(CE) + A(DE) = A(CDE), A(CD) + A(DE) = A(CDE), and A(CE) + A(DE) = A(CDE), as well as A(CD) + A(CE) + A(DE) = A(CDE). The same holds for the relationships of B. Hence, in each subset, one statement is redundant, reducing the total weight from 6 to 4.

It should be emphasized that elimination of redundant logically-dependent statements pertains to the data as a whole, as no particular statement is or should ever be excluded. All statements are relevant even if their value is less under certain circumstances.

To appreciate the effects of redundancy, consider again the example of multistate character A(B(CD)), which has four statements, A(BC), A(BD), A(CD) and B(CD). Its two derivative binary characters, A(BCD) and AB(CD), have five statements with A(CD) occurring twice. Hence one A(CD) state-ment is redundant. To remedy the situation, each of the three statements from A(BCD) can be downweighted by two-thirds, reducing the total informa-tion content of the two binary characters to four statements. Such down-weighting is referred to as 'fractional weighting' (FW). Fractional weighting highlights the difference between multistate characters and pairs of (con-gruent) binary characters in terms of their information content and informa-tion redundancy. In the multistate character A(B(CD)), all four statements are given equal weight (as states are assumed dependent). In the pair of binary characters, the statements for A(BCD) are dependent and hence have some logical redundancy but are still independent of AB(CD) (Table 9.2).

Three-item statement representation identifies non-independence in the data, whereas the standard approach does not. Redundancy thus remains with additive binary coding. Fractional weighting is therefore more sensitive

Table 9.2 Comparison of three item representation for the multistate character A(B(CD)), its binary equivalents and the effects of fractional weighting on both kinds of characters.

| | Uniform Weighting | | Fractional weighting | | Multistate |
	Binary A(BCD)	Binary AB(CD)	Binary A(BCD)	Binary AB(CD)	A(B(CD))
A(BC)	1	–	$\frac{2}{3}$	–	1
A(BD)	1	–	$\frac{2}{3}$	–	1
A(CD)	1	1	$\frac{2}{3}$	1	1
B(CD)	–	1	–	1	1
	3 + 2		2 + 2		4
Total	5		4		4

than uniform weighting to the information in the data. It should be noted that the use of the word 'weighting' refers to relative information content and should not be construed in the same terms as the 'weighting' procedures described in Chapter 5.

Consider once again the data in Table 9.1b. Three-item statements analysis using 'uniform weighting' (UW), in which all statements are given equal (unit) weight, yields one cladogram (Fig. 9.2a, length = 30; 18 ATS + 2 × 6 NTS = 30). Using fractional weighting, the optimal cladogram(s) will be that which has the greatest total weight, which need not be the same as the cladogram that accommodates the greatest number of statements (Nelson 1993, Nelson and Ladiges 1994). If fractional weighting is applied, using Hennig86 (which requires weights to have integer values between 0 and 10, see below), an additional cladogram is found (Fig. 9.2b). Both cladograms have a length of 27. The cladogram in Fig. 9.2a implies six three-item statements in various quantities, totalling 20 after weighting. This is four fewer than for the UW matrix (Table 9.3). Of these 20 statements, 15 are accommodated on the cladogram: A(BC) × 3, A(BD) × 3, A(CD) × 6 and B(CD) × 3. A further six statements are not accommodated: C(AB) × 3, D(AB) × 3 (= 6 statements). This gives a cladogram length of 27(= 15 + (2 × 6)). Likewise, for the cladogram in Fig. 9.2b, of the 20 three-item statements, 15 are accommodated: C(AB) × 3, D(AB) × 3, A(CD) × 6 and B(CD) × 3; and six statements are not: A(BC) × 3 and A(BD) × 3, also giving a length of 27.

To understand the fractional weights, inspection of the data is required. In Table 9.3, three statements contribute to the component A(BCD): A(BC), A(BD) and A(CD). Of these, only two are required to yield the correct result, the third being logically implied in each case. Therefore, a fractional weight of two-thirds is applied to each statement. Effectively, the FW matrix is reduced by one component A(BCD). In the UW matrix, A(BCD) is included

Table 9.3 Comparison of the effects of uniform weighting (UW) and fractional weighting (FW) using the data in Table 9.1b and cladograms in Fig. 9.2. Cladogram 1 = Fig. 9.2a; cladogram 2 = Fig. 9.2b. FW (no factor) = rounded fractional weight for use in Hennig86 or PAUP without a correction factor (the true value, if different, is given in parentheses). FW($\times 3$) = fractional weight to be applied following correction of the actual values by a factor of 3 to minimize the effects of rounding error, e.g. for statement A(BC), the original fractional weight of $2\frac{2}{3}$ is multiplied by 3 to give a new weight of 8. D = accommodated three-item statement (ATS) ($\checkmark$) or non-accommodated three item statement (NTS) ($\times$) on cladogram. Total = total statements from original data (column sums). Length = length of tree = sum of numbers of statements with a '$\checkmark$' in column D + 2 $\times$ statements with a '$\times$' in column D. See text for further discussion.

Statements	Cladogram 1				Statements	Cladogram 2			
	UW	FW (no factor)	FW ($\times 3$)	D		UW	FW (no factor)	FW ($\times 3$)	D
A(BC)	4	3 ($2\frac{2}{3}$)	8	$\checkmark$	A(BC)	4	3 ($2\frac{2}{3}$)	8	$\times$
A(BD)	4	3 ($2\frac{2}{3}$)	8	$\checkmark$	A(BD)	4	3 ($2\frac{2}{3}$)	8	$\times$
A(CD)	7	6 ($5\frac{2}{3}$)	17	$\checkmark$	A(CD)	7	6 ($5\frac{2}{3}$)	17	$\checkmark$
B(CD)	3	3	9	$\checkmark$	B(CD)	3	3	9	$\checkmark$
C(AB)	3	3	9	$\times$	C(AB)	3	3	9	$\checkmark$
D(AB)	3	3	9	$\times$	D(AB)	3	3	9	$\checkmark$
Total	24	20 (20)	60	4	Total	24	20 (20)	60	4
Length	30	27 (26)	78	4	Length	32	27 ($25\frac{1}{3}$)	76	4

as redundant statements, which constitutes spurious information.

This example is also instructive in another matter relating to implementation. Fractional weights are, by definition, fractions. However, all currently available parsimony programs can only apply integer weights to characters in a data matrix. Thus, the fractional weights must be rounded to the nearest integer value. In the above example, this results in two equally most parsimonious cladograms of length 27. However, if we carry out the exercise manually using the fractions themselves, then we arrive at a length of 26 for the cladogram in Fig. 9.2a and only $25\frac{1}{3}$ for that in Fig. 9.2b. We now see that there is, in fact, only one most parsimonious cladogram, not two (Table 9.3). The result of two equally most parsimonious solutions is an artefact of the implementation. To circumvent this problem, fractional weights should be multiplied by an appropriate correction factor to give accurate integer values. In the present example, this is achieved by using a factor of 3 (Table 9.3). In practice, however, rounding may still be required to give integer weights. For example, if a weight of $2\frac{2}{3}$ is interpreted as 2.667, then correction using a factor of 3 will give a weight of 8.001, which then needs to be rounded to 8.

9.3.6 Minimal cladograms

Analysis of a matrix of three-item statements may yield one or more cladograms, some of which may not be minimal. A minimal cladogram is understood here to be one that is not only minimal in length, but one in which all the resolved nodes are supported by data. (For a discussion of similar situations in standard analyses, see §4.2.) The use of question marks in any data matrix may create problems in resulting cladograms by leading to over-resolution of nodes not supported by data. Over-resolution of cladograms for three-item statements analysis can be illustrated graphically by considering one standard character. Out of seven taxa, six have the informative state, and thus there are 15 three-item statements. Analysis of these 15 statements using Hennig86 gives 945 solutions, each fully resolved and of length 15. This is because every statement is accommodated on every cladogram. Of the 945 solutions, there is none in which all nodes (of which there are five) are fully supported by data. The strict consensus tree of the 945 cladograms is also a minimal cladogram of length 15 but has only one node supported by data, the node uniting all six taxa with the informative state. This is the preferred cladogram, not only because it is of minimum length, but also because it includes no nodes that are unsupported by data. This observation applies to all single binary characters represented as three-item statements where $n > 2$ (Nelson and Ladiges 1992). This is the reason why analysis of the matrix in Fig. 9.1f by Hennig86 results in three cladograms, of which the strict consensus is the preferred tree.

In many three-item statements analyses, the minimal cladogram can be found by constructing the strict consensus tree of all the most parsimonious cladograms, *providing* it is the same length as those cladograms (the strict consensus tree collapses nodes not supported by data). When the strict consensus tree is longer than all the most parsimonious cladograms, it cannot be minimal. In cases where the strict consensus tree is not minimal, one may inspect each most parsimonious cladogram, collapsing nodes manually, and noting any change in length. If a node can be collapsed with no change in cladogram length, then the resultant less-resolved topology is considered preferable, because originally the collapsed node had no support from the data.

An alternative strategy is to use the parsimony program, NONA, which eliminates ambiguous optimizations due to the use of question marks. Results so far indicate that NONA usually, but not always, finds the minimal cladogram(s) for a three-item statements matrix more efficiently than manual manipulation.

9.3.7 Optimization

Because of their design, current parsimony programs may indicate ambiguous optimizations for some statements at some nodes on a cladogram. The programs treat every statement (every column in the matrix) as a conventional character. That is, they must assign a value to each character on every node of the cladogram despite the statement having only three real values. Consider the example in Fig. 9.3. If the statement A(CD) is optimized onto the cladogram A(B(C(DE))), the programs will assign 0/1 to node X and 1 to node Y (Fig. 9.3a). The assignment of 0/1 to node X is irrelevant to the understanding of the statement's purpose. All that matters is that the statement fits node Y exactly (Fig. 9.3b). Note that a three-item statement is read on a cladogram in the form 'C and D are more closely related to each

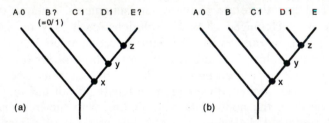

(a)   (b)

Fig. 9.3 'Optimization' of the three-item statement A(CD) onto the cladogram A(B(C(DE))). (a) Parsimony computer programs that require all cells of a data matrix to be filled and attempt to optimize the 'condition' in taxon B as either 0 or 1, even though this is nonsensical. (b) The correct 'optimization' of the three-item statement A(CD), in which taxon B is irrelevant. The statement A(CD) should be read as 'C and D are more closely related to each other than either is to A' and not in the form of the standard approach, which sees nodes as possible (ancestral) 'transformations' of one character (or state) into another.

other than either is to A' not as in the standard approach, which sees nodes as possible (ancestral) transformations of one character (or state) into another. This point was misunderstood by Farris *et al.* (1995) who, in their examples, count three-item statements as if they are optimized characters. The assignment at node X in Fig. 9.3a is due to the programs treating question marks as 'potential' data when, in the case of three-item statements, they are no such thing (the standard approach also has problems; see §4.2). The main point is that despite the default optimization of values, 'Hennig86 (and PAUP) efficiently implements three-item analysis because tree length, if not optimization, is *exact*' (Nelson 1993) (our emphasis). Cladogram length still reflects optimality accurately.

9.3.8 Information measures: CI and RI

A three-item statement involves only three terminals and therefore will either fit to a node on a cladogram or not. In other words, a three-item statement will display either one step and have a ci of 1.00, or two steps and have a ci of 0.50. Hence CI is not a useful measure as it simply distinguishes the fit to a node of each statement.

For ri values, the situation is somewhat different. Platnick (1993) drew attention to how three-item statements differ from binary characters in their fit to cladograms, by noting the performance of a suite of three-item statements on a series of different (specified) topologies (Table 9.4) The topologies of the cladograms are not relevant to this discussion. Note that RI reflects the number of accepted (accommodated) three-item statements as a fraction of the total number of three-item statements considered. For example, cladogram 1 accepts 65 statements as true, hence $65/135 = 0.48$; cladograms 2 and 3 accept 54 statements as true, hence $54/135 = 0.40$; cladogram 4 accepts no statements as true, hence $0/135 = 0$; and cladogram 5 accepts all 135 statements as true, hence $135/135 = 1$. Therefore RI would appear to be a useful measure for the amount of fit for each possible cladogram.

Table 9.4 Data for five cladograms discussed by Platnick (1993). Each cladogram is tabulated for the number of accepted (accommodated) and prohibited (non-accommodated) statements, together with some more conventional statistics.

	Number of Statements					
	Accepts ('True')	Prohibits ('False')	Steps	Nodes	Length	RI
Cladogram 1	65	70	6	12	205	48
Cladogram 2	54	81	2	11	216	40
Cladogram 3	54	81	2	4	216	40
Cladogram 4	0	135	6	6	270	0
Cladogram 5	135	0	1	1	135	100

9.3.9 Summary of implementation procedures

Implementation of three-item statements analysis can be executed in a manner similar to standard character analysis but requires the same attention to detail. This is because all currently available parsimony programs were designed with a different purpose in mind. Nevertheless, to repeat, 'Hennig86 (and PAUP) efficiently implements three-item analysis because tree length, if not optimization, is exact' (Nelson 1993).

Improvements in precision in three-item statements analysis come from three sources. First, attention must be given to redundancy in the data and fractional weights should be used. Second, the scaling of weights must be appropriate to avoid oversimplifying the effect of integer weights currently required by parsimony programs. Third, the final cladogram(s) should be minimal with respect first to length and then to the number of nodes supported by data.

Platnick *et al.* (1991*a*) suggested that the best cladogram for available data should satisfy 'the criteria of parsimony, relative informativeness of characters, and maximum resolution of characters'. In both three-item statements analysis and the standard approach, parsimony is the principle used to fit data to a cladogram and fully resolved cladograms are the objective. It would seem that 'maximum resolution of characters' is an issue that has only recently been explored. The relative informativeness of characters differs depending on the way data are represented, and this may be the major distinguishing factor between the standard and three-item approaches (see also §4.2).

9.4 PRECISION

When Nelson and Platnick (1991) first proposed the use of three-item statements to analyse systematic data, they suggested that it might improve the precision of parsimony (*sensu* Farris 1983). They presented results from the analysis of several hypothetical and one real data matrix (from Carpenter 1988) contrasting the results of three-item and standard analyses. Their results showed that three-item statements analysis sometimes produced fewer cladograms, sometimes more cladograms, and sometimes different cladograms compared with the standard approach. What, then, is the meaning of 'more precise'?

The suggestions of Platnick *et al.* (1991*a*) once again provide a valuable means of understanding the issues facing cladistic practice. The best cladogram for available data should satisfy 'the criteria of parsimony, relative informativeness of characters, and maximum resolution of taxa'. As stated above, both three-item statements analysis and the standard approach use parsimony to organise the data. Both approaches also attempt to gain 'maximum resolution of taxa', such that all nodes are truly supported by data.

Table 9.5 (a) Matrix of conflicting characters for four taxa (A–D) and an all-zero root (O), coded for three characters 1–3. (b) The equivalent three-item statements matrix. (After Nelson 1996.)

(a)

	1	2	3
O	0	0	0
A	0	0	0
B	0	1	1
C	1	0	1
D	1	1	0

(b)

	1 a	1 b	2 a	2 b	3 a	3 b
O	0	0	0	0	0	0
A	0	?	0	?	0	?
B	?	0	1	1	1	1
C	1	1	?	0	1	1
D	1	1	1	1	?	0

The only factor that can differ between the two approaches is the 'relative informativeness of characters' or, perhaps more accurately, the relative informativeness of observations.

Nelson (1996) provided a series of examples that demonstrated improved precision by greater resolution of character data. Nelson analysed a series of data sets with four to seven taxa. Taxon A has all plesiomorphic states (0s), while the other taxa, B − n (where n = CD, CDE, up to CDEFGHI), have different combinations of conflicting apomorphic states (1s). Each matrix was analysed with an all-zero root, O, as the focus of interest was the relationships among taxa A − n. Taxon A is not an outgroup but part of the problem requiring solution.

For example, standard analysis of the data in Table 9.5a yields six equally most parsimonious cladograms, the strict consensus tree of which is an uninformative bush. This result suggests that there is no overall information in this matrix—or at least the information has maximum conflict. If the same data are represented as a suite of three-item statements (Table 9.5b), three cladograms result, the strict consensus tree of which is A(BCD). This result suggests that there is information in this matrix to relate B + C + D as a group relative to A.

Nelson (1996) analysed 120 matrices, in which there were conflicting characters in three (BCD) to eight (B–I) taxa, using the standard approach (although the series could be extended indefinitely). The results from 96 of these matrices included the group B − n relative to A, while 24 did not. In other words, over three-quarters resolve a node, which is sufficient to suspect that matrices of this kind are generally informative. Three-item statements analysis of the same 120 matrices *always* yields a resolved node.

A glance at the data in Table 9.5 seems to imply that while the relationships among B–D cannot be specified (there is significant conflict), they do indeed share a closer relationship to each other than they do to taxon A. Three-item statements analysis does seem to improve precision in cases with this kind of conflict (see also Siebert and Williams 1997). Thus, it would seem that for the standard approach, a taxon that lacks the attribute of a larger group (B–D), while retaining information relevant to relationships within the group, collapses the cladogram entirely (a point noted earlier by Nelson and Platnick 1980). This suggests that in order to understand the relationships of a subgroup, one needs to have established the relationships of the larger group (Nelson 1996). Finally, these results suggest that the behaviour of the standard approach to certain kinds of conflict might be a consequence of a particular 'model' of character evolution underlying its implementation, a model 'that requires synapomorphy (evidence of relationship) to have a unique origin (optimized as 1 at a node with distal 0s as reversal(s), or vice versa)' (Nelson 1992; also 1996).

Platnick *et al.* (1996) suggested a second demonstration of increased precision. Using a simple example of one binary character, distributed among four taxa (A–D) such that C and D share the apomorphic state and A and B share the plesiomorphic state, standard analysis yields the cladogram AB(CD). Platnick *et al.* (1996) suggested that inspection of all 26 possible cladograms for the four taxa (Fig. 9.4) would be instructive, because this would give an indication of how much 'worse' the others were compared with the correct solution. When the binary character AB(CD) is fitted to the cladograms, these can be divided into two series. Four cladograms (Fig. 9.4a–c and y) have one step, while the remaining 22 have two steps. Examination of the cladograms with two steps implies that, among other things, completely incorrect solutions (e.g. Fig. 9.4g–j or Fig. 9.4l–o) are as good (or as bad) as some partially correct solutions (e.g. Fig 9.4d–e) or even the totally unresolved bush (Fig. 9.4z). This point had been made earlier by Platnick (1989: 23) in a different context.

The binary character AB(CD) yields two three-item statements, A(CD) and B(CD) and three-item statements analysis yields the cladogram AB(CD). If these statements are also fitted to all 26 possible solutions, three series are obtained, rather than two. The three-item statements solutions comprise four cladograms with two steps (Fig. 9.4a–c and y), six with 3 steps (Fig. 9.4d–f, k, r–s) and 16 with 4 steps. In this context, three-item statements analysis is more 'precise', in the sense that it partitions all the possible solutions in a more efficient way than does the standard approach.

What, then, is the significance of such partitioning, given that all cladograms with a length of 3 are sub-optimal relative to the most parsimonious, minimal cladogram (Fig. 9.4y)? The significance lies in the value of different cladograms as more data accumulate. For example, using the standard

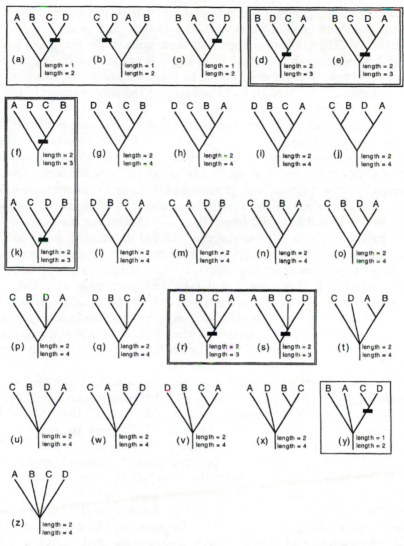

Fig. 9.4 (a–z) The 26 possible topologies for four taxa, A–D. Two lengths are given for each cladogram. The upper figure relates to the binary character, AB(CD); the lower to its two constituent three-item statements, A(CD) and B(CD). The four cladograms enclosed within single-line boxes (a–c, y) accommodate the binary character and both three-item statements with a single step each and thus explain all of the data. The binary character fits to all other cladograms with two steps, which thus appear to explain none of the data. However, the six cladograms enclosed within double-line boxes (d–f, k, r–s) do accommodate one of the three-item statements with one step and therefore explain at least part of the data. Thus, although suboptimal, they may be preferred to all of other remaining cladograms (g–j, l–q, t–x, z), which require two steps to accommodate each three-item statement.

approach, the cladograms in Fig. 9.4k and Fig. 9.4l explain none of the data. However, using three-item statements, the former cladogram includes one statement, A(CD), and hence explains at least some of the data. As Platnick *et al.* (1996) pointed out, switching from Fig. 9.4l to Fig. 9.4k should not be a 'zero-cost' option. The value of the three-item statements approach in this context is that it is likely to be more sensitive to the accumulation of further data than is the standard approach.

There is a further point that seems significant for coding protocols. The standard approach treats the binary character AB(CD) as a feature relating C and D together with respect to A *and* B. It says nothing about the relationships of C and D relative to A *or* B. The standard character is restricted to treating C + D as a group, with plesiomorphic states as uninformative with respect to relationships. Under certain circumstances, the standard approach will treat A + B as the 'group' rather than C + D, with the plesiomorphy (0) interpreted as 'secondary' or a 'reversal'. This is different from crude phenetic grouping by symplesiomorphy but nevertheless constitutes a version of 'grouping by plesiomorphy' (de Pinna 1996). Platnick *et al.* (1996) went further, wondering 'how long it will take for systematists to realise that allowing the '0' entries for taxa A and B to constrain potential resolutions can also give this sort of 'negative evidence' more weight than it deserves, in at least some circumstances'. It would also seem that the notion of character reversal belongs to the realm of 'trees' rather than cladograms and thus again constitutes a kind of model (see Chapter 1).

Of further interest is that the fully resolved solutions selected by the standard approach as shortest (Fig. 9.4a–c) correspond to the 'Interpretation 1' solutions, originally proposed by Nelson and Platnick (1980) for dealing with potential resolution of basal trichotomies. In contrast, the fully resolved solutions selected by the three-item statements approach as shortest (Fig. 9.4a–f and k) correspond to the 'Interpretation 2' solutions proposed by Nelson and Platnick (1980), where the close relationship of C and D is maintained, even though A or B are more closely related to either C or D. (The additional solutions with terminal trichotomies simply represent summaries of the resolved cladograms. The cladogram in Fig. 9.4r is a summary of the fully resolved cladograms in Fig. 9.4c–e, and the cladogram in Fig. 9.4s is a summary of the fully resolved cladograms in Fig. 9.4a, f and k.) Finally, Interpretations 1 and 2 bear some resemblance to Assumptions 1 and 2 in biogeography, while 'secondary' symplesiomorphy, 'reversals', and plesiomorphies as 'potentially informative' have a certain amount of similarity with Assumption 0 of biogeography, a somewhat questionable protocol (Humphries and Parenti in press).

It is worth recalling that three-item statements analysis began in biogeography. Biogeography, in its modern cladistic form, deals with cladograms of areas (for a recent summary, see Humphries and Parenti in press). A more general question with respect to cladistics might be: is there an empirical way of dealing with all branching diagrams, regardless of the 'kind' of data? Is there a general theory of cladograms and hence a general theory of systematics? The first analytical explorations of this question began, perhaps, with Nelson and Platnick (1981), a much neglected but still highly relevant and fundamental book. We may possibly question any direct analogy between systematics and biogeography. However, one understanding of the differences seems to reside in how characters are viewed, and one possible resolution may lie in rejecting characters as 'transformation series' (ancestor–descendent sequences) in favour of characters as statements of relationship (A is more closely related to B than it is to C).

9.5 CHAPTER SUMMARY

1. 'Cladistics' is grouping by synapomorphy, where synapomorphy is understood as evidence of relationship. Three-item statements analysis is an alternative way to code data based on the idea that 'taxon' and 'homology' represent the same relationship.

2. Three-item statements analysis reduces data to their simplest expression of relationship, the three-item statement. With four taxa (A–D), where C and D possess a particular feature and A and B do not, there are two three-item statements: A(CD) and B(CD).

3. A matrix of three-item statements can be analysed using current parsimony programs to find the best fitting cladogram.

4. The relationship between cladogram length and data (the suite of three-item statements) is given by: length = ATS (accommodated three-item statements) + 2 × NTS (non-accommodated three-item statements). Length accurately reflects the 'best' cladogram because 'Hennig86 (and PAUP) efficiently implements three-item statements analysis because tree length, if not optimization, is *exact*' (Nelson 1993).

5. Some binary characters may have redundant information. For example, the character representing the relationship A(BCD), has three statements, A(BC), A(BD) and A(CD), any pair of which in combination will yield the same result and thus logically imply the third. This suggests that in some cases not all statements are independent. In this example, as any two of the three statements derived from A(BCD) produce the same result, each can be weighted accordingly. Such downweighting is called fractional weighting.

6. Analysis of a matrix of three-item statements may yield one or more cladograms, some of which may not be minimal. A minimal cladogram is understood as one that is of minimal length and has all resolved nodes supported by data.

7. A statement should be read from a cladogram in the form 'C and D are more closely related to each other than either is to A', and not in the form of the standard approach, which sees nodes as possible (ancestral) 'transformations' of one character (or state) into another. Standard optimization is irrelevant.

8. When fitted to a cladogram, a statement will have a ci of either 1.00 (when a statement fits a particular cladogram) or 0.50 (when it does not). Hence CI is not a useful overall measure of fit as it simply distinguishes the fit to a node of each statement.

9. The number of accepted (accommodated) three-item statements as a fraction of the total number of statements considered is reflected by the RI value. Therefore, RI might be a useful measure for the amount of fit for each possible cladogram.

10. To improve precision in three-item statements analysis, attention must be given to redundancy in the data and fractional weights used. Weights should be scaled appropriately, in order to avoid introducing errors due to requirement of current computer programs for integer weights. The final cladogram(s) should be minimal with all nodes supported by data.

11. One understanding of the difference between the standard approach and the three-item approach resides in how character data are viewed: as 'transformation series' (ancestor–descendent sequences) or as statements of relationships (A is more closely related to B than it is to C).

References

References cited

Adams, E. N. (1972). Consensus techniques and the comparison of taxonomic trees. *Systematic Zoology*, **21**, 390–7.

Alberch, P. (1985). Problems with the interpretation of developmental sequences. *Systematic Zoology*, **34**, 46–58.

Allard, M. W. and Carpenter, J. M. (1996). On weighting and congruence. *Cladistics*, **12**, 183–98.

Almeida, M. T. and Bisby, F. A. (1984). A simple method for establishing taxonomic characters from measurement data. *Taxon*, **33**, 405–9.

Anderberg, A. and Tehler, A. (1990). Consensus trees, a necessity in taxonomic practice. *Cladistics*, **6**, 399–402.

Archie, J. W. (1985). Methods for coding variable morphological features for numerical taxonomic analysis. *Systematic Zoology*, **34**, 326–45.

Archie, J. W. and Felsenstein, J. (1993). The number of evolutionary steps on random and minimum length trees for random evolutionary data. *Journal of Theoretical Biology*, **45**, 52–79.

Baer, K. E. von (1828). *Ueber Entwickelungsgeschichte der Thiere: Beobachtung und Reflexion,* Theil 1. Gebrüder Bornträger, Königsberg.

Barthélemy, J.-P. and McMorris, F. R. (1986). The median procedure for n-trees. *Journal of Classification*, **3**, 329–34.

Barthélemy, J.-P. and Monjardet, B. (1981). The median procedure in cluster analysis and social choice theory. *Mathematical Social Sciences*, **1**, 235–67.

Baum, B. R. (1988). A simple procedure for establishing discrete characters from measurement data, applicable to cladistics. *Taxon*, **37**, 63–70.

Begle, D. P. (1991). Relationships of the osmeroid fishes and the use of reductive characters in phylogenetic analysis. *Systematic Zoology*, **40**, 33–53.

Bremer, K. (1988). The limits of amino-acid sequence data in angiosperm phylogenetic reconstruction. *Evolution*, **42**, 795–803.

Bremer, K. (1990). Combinable component consensus. *Cladistics*, **6**, 369–72.

Bremer, K. (1994). Branch support and tree stability. *Cladistics*, **10**, 295–304.

Brower, A. V. Z. and Schawaroch, V. (1996). Three steps to homology assessment. *Cladistics*, **12**, 265–72.

Bryant, H. N. (1992). The role of permutation tail probability tests in phylogenetic systematics. *Systematic Biology*, **41**, 258–63.

Bryant, H. N. (1995). Why autapomorphies should be removed: a reply to Yeates. *Cladistics*, **11**, 381–4.

Bull, J. J., Huelsenbeck, J. P., Cunningham, C. W., Swofford, D. L. and Waddell, P. J. (1993). Partitioning and combining data in phylogenetic analysis. *Systematic Biology*, **42**, 384–97.

Cain, A. J. and Harrison, G. A. (1958). An analysis of the taxonomist's judgement of affinity. *Proceedings of the Zoological Society of London*, **131**, 85–98.

Camin, J. H. and Sokal, R. R. (1965). A method for deducing branching sequences in phylogeny. *Evolution*, **19**, 311–26.

Carpenter, J. M. (1988). Choosing among multiple equally parsimonious cladograms. *Cladistics*, **4**, 291–6.

Carpenter, J. M. (1992). Random cladistics. *Cladistics*, **8**, 147–53.

Carpenter, J. M. (1996). Uninformative bootstrapping. *Cladistics*, **12**, 177–81.

Chappill, J. A. (1989). Quantitative characters in phylogenetic analysis. *Cladistics*, **5**, 217–34.

Coddington, J. and Scharff, N. (1994). Problems with zero-length branches. *Cladistics*, **10**, 415–23.

Colless, D. H. (1980). Congruence between morphological and allozyme data for *Menidia* species: a reappraisal. *Systematic Zoology*, **29**, 288–99.

Cranston, P. S. and Humphries, C. J. (1988). Cladistics and computers: a chironomid conundrum. *Cladistics*, **4**, 72–92.

Davis, J. I. (1993). Character removal as a means for assessing the stability of clades. *Cladistics*, **9**, 201–10.

DeBry, R. W. and Slade, N. A. (1985). Cladistic analysis of restriction endonuclease cleavage maps within a maximum-likelihood framework. *Systematic Zoology*, **34**, 21–34.

de Queiroz, K. (1985). The ontogenetic method for determining character polarity and its relevance to phylogenetic systematics. *Systematic Zoology*, **34**, 280–99.

de Queiroz, K. (1993). For consensus (sometimes). *Systematic Biology*, **42**, 368–72.

Devereux, R., Loehlich, A. R. and Fox, G. E. (1990). Higher plant origins and the phylogeny of green algae. *Journal of Molecular Evolution*, **31**, 18–24.

Dixon, M. T. and Hillis, D. M. (1993). Ribosomal RNA secondary structure: compensatory mutations and implications for phylogenetic analysis. *Molecular Biology and Evolution*, **10**, 256–67.

Donoghue, M. J., Olmstead, R. G., Smith, J. F. and Palmer, J. D. (1992). Phylogenetic relationships of Dipsacales based on *rbc*L sequences. *Annals of the Missouri Botanical Garden*, **79**, 672–85.

Eernisse, D. J. and Kluge, A. G. (1993). Taxonomic congruence versus total evidence, and amniote phylogeny inferred from fossils, molecules, and morphology. *Molecular Biology and Evolution*, **10**, 1170–95.

Eldredge, N. (1979). Alternative approaches to evolutionary theory. *Bulletin of the Carnegie Museum of Natural History*, **13**, 7–19.

Faith, D. P. (1991). Cladistic permutation tests for monophyly and non-monophyly. *Systematic Zoology*, **40**, 366–75.

Faith, D. P. and Cranston, P. S. (1991). Could a cladogram this short have arisen by chance alone? - On permutation tests for cladistic structure. *Cladistics*, **7**, 1–28.

Farris, J. S. (1969). A successive approximations approach to character weighting. *Systematic Zoology*, **18**, 374–85.

Farris, J. S. (1970). Methods for computing Wagner trees. *Systematic Zoology*, **19**, 83–92.

Farris, J. S. (1971). The hypothesis of nonspecificity and taxonomic congruence. *Annual Review of Ecology and Systematics*, **2**, 277–302.

Farris, J. S. (1972). Estimating phylogenetic trees from distance matrices. *American Naturalist*, **106**, 645–68.

Farris, J. S. (1983). The logical basis of phylogenetic analysis. *Advances in Cladistics*, **2**, 1–36.

Farris, J. S. (1988). *Hennig86, version 1.5*. Program and documentation. Port Jefferson Station, New York.

Farris, J. S. (1989). The retention index and the rescaled consistency index. *Cladistics*, **5**, 417–19.

Farris, J. S., Kluge, A. G. and Eckhardt, M. J. (1970). A numerical approach to phylogenetic systematics. *Systematic Zoology*, **19**, 172–89.

Farris, J. S., Källersjö, M., Kluge, A. G. and Bult, C. (1994). Testing significance of incongruence. *Cladistics*, **10**, 315–19.

Farris, J. S., Källersjö, M., Albert, V. A., Allard, M., Anderberg, A., Bowditch, B., Bult, C., Carpenter, J. M., Crow, T. M., De Laet, J., Fitzhugh, K., Frost, D., Goloboff, P. A., Humphries, C. J., Jondelius, U., Judd, D., Karis, P. O., Lipscomb, D., Luckow, M., Mindell, D., Muona, J., Nixon, K. C., Presch, W., Seberg, O., Siddall, M. E., Struwe, L., Tehler, A., Wenzel, J., Wheeler, Q. D. and Wheeler, W. (1995). Explanation. *Cladistics*, **11**, 211–8.

Finden, C. R. and Gorden, A. D. (1985). Obtaining common pruned trees. *Journal of Classification*, **2**, 255–76.

Fitch, W. M. (1971). Towards defining the course of evolution: minimum change for a specified tree topology. *Systematic Zoology*, **20**, 406–16.

Fitch, W. M. and Ye, J. (1990). Weighted parsimony: does it work? In *Phylogenetic analysis of DNA sequences* (ed. M. M. Miyamoto and J. Cracraft), pp. 147–54. Oxford University Press, Oxford.

Gaffney, E. S., Meylan, P. A. and Wyss, A. R. (1991). A computer assisted analysis of the relationships of the higher categories of turtles. *Cladistics*, **7**, 313–35.

Gardiner, B. (1993). Haematothermia: warm-blooded amniotes. *Cladistics*, **9**, 369–95.

Gauthier, J., Kluge, A. G. and Rowe, T. (1988). Amniote phylogeny and the importance of fossils. *Cladistics*, **4**, 105–209.

Goldman, N. (1988). Methods for discrete coding of morphological characters in numerical analysis. *Cladistics*, **4**, 59–71.

Goloboff, P. A. (1991). Homoplasy and the choice among cladograms. *Cladistics*, **7**, 215–32.

Goloboff, P. A. (1993). Estimating character weights during tree search. *Cladistics*, **9**, 83–91.

Goloboff, P. A. (1995a). Parsimony and weighting: a reply to Turner and Zandee. *Cladistics*, **11**, 91–104.

Goloboff, P. A. (1995b). A revision of the South American spiders of the family Nemesiidae (Araneae, Mygalomorphae). Part I. Species from Peru, Chile, Argentina, and Uruguay. *Bulletin of the American Museum of Natural History*, **224**, 1–189.

Goloboff, P. A. (1996). Methods for faster parsimony analysis. *Cladistics*, **12**, 199–220.

Grande, L. (1994). Repeating patterns in nature, predictability, and 'impact' in science. In *Interpreting the hierarchy of nature* (ed. L. Grande and O. Rieppel), pp. 61–84. Academic Press, San Diego.

Harshman, J. (1994). The effect of irrelevant characters on bootstrap values. *Systematic Biology*, **43**, 419–24.

Harvey, A. W. (1992). Three-taxon statements: more precisely, an abuse of parsimony? *Cladistics*, **8**, 345–54.

Hedges, S. B. and Maxson, L. R. (1996). Re: molecules and morphology in amniote phylogeny. *Molecular Phylogenetics and Evolution*, **6**, 312–14.

Hedges, S. B., Moberg, K. D. and Maxson, L. R. (1990). Tetrapod phylogeny inferred from 18S and 28S ribosomal RNA sequences and a review of the evidence for amniote relationships. *Molecular Biology and Evolution*, **7**, 607–33.

Hendy, M. D., Little, C. H. C. and Penny, D. (1984). Comparing trees with pendent vertices labelled. *SIAM Journal of Applied Mathematics*, **44**, 1054–65.

Hennig, W. (1950). *Grundzüge einer Theorie der phylogenetischen Systematik*. Deutsche Zentralverlag, Berlin.

Hennig, W. (1965). Phylogenetic systematics. *Annual Review of Entomology*, **10**, 97–116.

Hennig, W. (1966). *Phylogenetic systematics*. University of Illinois Press, Urbana.

Hillis, D. M. (1991). Discriminating between phylogenetic signal and random noise in DNA sequences. In *Phylogenetic analysis of DNA sequences* (ed. M. M. Miyamoto and J. Cracraft), pp. 278–94. Oxford University Press, Oxford.

Hillis, D. M., Allard, M. W. and Miyamoto, M. M. (1993). Analysis of DNA sequence data: phylogenetic inference. In *Molecular Evolution: producing the biochemical data*. Methods in Enzymology, No. 224, (ed. E. A. Zimmer, T. J. White, R. L. Cann and A. C. Wilson), pp. 456–87. Academic Press, San Diego.

Huelsenbeck, J. P. (1991a). Tree-length skewness: an indicator of phylogenetic information. *Systematic Zoology*, **40**, 257–70.

Huelsenbeck, J. P. (1991b). When are fossils better than extant taxa in phylogenetic analysis? *Systematic Biology*, **40**, 458–69.

Huelsenbeck, J. P. and Bull, J. J. (1996). A likelihood ratio for detection of phylogenetic signal. *Systematic Biology*, **45**, 92–8.

Huelsenbeck, J. P., Bull, J. J. and Cunningham, C. W. (1996). Combining data in phylogenetic analysis. *Trends in Ecology and Systematics*, **11**, 152–8.

Huey, R. B. and Bennett, A. F. (1987). Phylogenetic studies of co-adaptation: preferred temperatures versus optimal performance temperatures of lizards. *Evolution*, **41**, 1098–115.

Humphries, C. J. and Parenti, L. In press. *Cladistic Biogeography*. (2nd ed.). Clarendon Press, Oxford.

Jardine, N. (1969). A logical basis for biological classification. *Systematic Zoology*, **18**, 37–52.

Källersjö, M., Farris, J. S., Kluge, A. G. and Bult, C. (1992). Skewness and permutation. *Cladistics*, **8**, 275–87.

Kluge, A. G. (1985). Ontogeny and phylogenetic systematics. *Cladistics*, **1**, 13–27.

Kluge, A. G. (1988). The characteristics of ontogeny. In *Ontogeny and systematics* (ed. C. J. Humphries), pp. 57–82. Columbia University Press, New York.

Kluge, A. G. (1989). A concern for evidence and a phylogenetic hypothesis of relationships among *Epicrates* (Boidae: Serpentes). *Systematic Zoology*, **38**, 7–25.

Kluge, A. G. (1993). Three-taxon transformation in phylogenetic inference: ambiguity and distortion as regards explanatory power. *Cladistics*, **9**, 246–59.

Kluge, A. G. (1994). Moving targets and shell games. *Cladistics*, **10**, 403–13.

Kluge, A. G. and Strauss, R. E. (1985). Ontogeny and systematics. *Annual Review of Ecology and Systematics*, **16**, 247–68.

Kluge, A. G. and Wolf, J. (1993). Cladistics: what's in a word? *Cladistics*, **9**, 1–25.

Kraus, F. (1988). An empirical evaluation of the use of the ontogeny polarization criterion in phylogenetic inference. *Systematic Zoology*, **37**, 106–41.

Kubicka, E., Kubicka, G. and McMorris, F. R. (1995). An algorithm to find agreement subtrees. *Journal of Classification*, **12**, 91–9.

Laconte, H. and Stevenson, D. W. (1991). Cladistics of the Magnoliidae. *Cladistics*, **7**, 267–96.

Lanyon, S. M. (1985). Detecting internal inconsistencies in distance data. *Systematic Zoology*, **34**, 397–403.

Larson, A. (1994). The comparison of morphological and molecular data in phylogenetic systematics. In *Molecular approaches to ecology and evolution* (ed. B. Schierwater, B. Streit, G. P. Wagner and R. DeSalle), pp. 371–90. Birkhäuser Verlag, Basel.

Lauder, G. V. (1990). Functional morphology and systematics: studying functional patterns in an historical context. *Annual Review of Ecology and Systematics*, **21**, 317–40.

Littlewood, D. T. J. and Smith, A. B. (1995). A combined morphological and molecular phylogeny for sea urchins (Echinoidea: Echinodermata). *Philosophical Transactions of the Royal Society of London*, B, **347**, 213–34.

Løvtrup, S. (1978). On von Baerian and Haeckelian recapitulation. *Systematic Zoology*, **27**, 348–52.

Maddison, W. P. (1993). Missing data versus missing characters in phylogenetic analysis. *Systematic Biology*, **42**, 576–81.

Maddison, W. P., Donoghue, M. J. and Maddison, D. R. (1984). Outgroup analysis and parsimony. *Systematic Zoology*, **33**, 83–103.

Margush, T. and McMorris, F. R. (1981). Consensus n-trees. *Bulletin of Mathematical Biology*, **43**, 239–44.

Marshall, C. (1992). Substitution bias, weighted parsimony, and amniote phylogeny as inferred from 18S rRNA sequences. *Molecular Biology and Evolution*, **9**, 370–73.

Mayr, E. (1969). *Principles of systematic zoology*. McGraw-Hill, New York.

Mayr, E., Linsley, E. G. and Usinger, R. L. (1953). *Methods and principles of systematic zoology*. McGraw-Hill, New York.

Meier, R. (1994). On the inappropriateness of presence/absence recoding for non-additive multistate characters in computerized cladistic analyses. *Zoologischer Anzeiger*, **232**, 201–12.

Mickevich, M. F. (1982). Transformation series analysis. *Systematic Zoology*, **31**, 461–78.

Mickevich, M. F. and Johnson, M. F. (1976). Congruence between morphological and allozyme data in evolutionary inference and character evolution. *Systematic Zoology*, **25**, 260–70.

Mishler, B. (1994). Cladistic analysis of molecular and morphological data. *American Journal of Physical Anthropology*, **94**, 143–56.

Mitter, C. (1980). The Thirteenth Annual Numerical Taxonomy Conference. *Systematic Zoology*, **29**, 177–90.

Miyamoto, M. M. (1985). Consensus cladograms and general classification. *Cladistics*, **1**, 186–9.

Miyamoto, M. M. and Boyle, S. M. (1989). The potential importance of mitochondrial DNA sequence data to eutherian mammal phylogeny. In *The hierarchy of life* (ed.

B. Fernholm, K. Bremer and H. Jornvall), pp. 437–50. Elsevier Science, Amsterdam.

Miyamoto, M. M. and Fitch, W. M. (1995). Testing species phylogenies and phylogenetic methods with congruence. *Systematic Biology*, **44**, 64–7.

Mueller, L. D. and Ayala, F. J. (1982). Estimation and interpretation of genetic distance in empirical studies. *Genetical Research*, **40**, 127–37.

Neff, N. (1986). A rational basis for a priori character weighting. *Systematic Zoology*, **35**, 110–23.

Nelson, G. J. (1973). The higher-level phylogeny of the vertebrates. *Systematic Zoology*, **22**, 87–91.

Nelson, G. J. (1978). Ontogeny, phylogeny, paleontology, and the biogenetic law. *Systematic Zoology*, **27**, 324–45.

Nelson, G. J. (1979). Cladistic analysis and synthesis: principles and definitions, with a historical note on Adanson's Famille des Plantes (1763–1764). *Systematic Zoology*, **28**, 1–21.

Nelson, G. J. (1992). Reply to Harvey. *Cladistics*, **8**, 355–60.

Nelson, G. J. (1993). Reply. *Cladistics*, **9**, 261–65.

Nelson, G. J. (1994). Homology and systematics. In *Homology: the hierarchical basis of comparative biology* (ed. B. K. Hall), pp. 101–49. Academic Press, San Diego.

Nelson, G. J. (1996). Nullius in verba. *Journal of Comparative Biology*, **1**, 141–52.

Nelson, G. J. and Ladiges, P.Y. (1992). Information content and fractional weight of three-taxon statements. *Systematic Biology*, **41**, 490–4.

Nelson, G. J. and Ladiges, P.Y. (1993). Missing data and three-item analysis. *Cladistics*, **9**, 111–13.

Nelson, G. J. and Ladiges, P.Y. (1994). Three-item consensus: empirical test of fractional weighting. In *Models in phylogeny reconstruction*, Systematics Association Special Volume, No. 52, (ed. R. W. Scotland, D. J. Siebert and D. M. Williams), pp. 193–207. Clarendon Press, Oxford.

Nelson, G. J. and Patterson, C. (1993). Cladistics, sociology and success: a comment on Donoghue's critique of David Hull. *Biology and Philosophy*, **8**, 441–3.

Nelson, G. J. and Platnick, N. I. (1980). Multiple branching in cladograms: two interpretations. *Systematic Zoology*, **29**, 86–91.

Nelson, G. J. and Platnick, N. I. (1981). *Systematics and biogeography: cladistics and vicariance*. Columbia University Press, New York.

Nelson, G. J. and Platnick, N. I. (1991). Three-taxon statements: a more precise use of parsimony? *Cladistics*, **7**, 351–66.

Nixon, K. C. and Carpenter, J. M. (1993). On outgroups. *Cladistics*, **9**, 413–26.

Nixon, K. C. and Carpenter, J. M. (1996*a*). On simultaneous analysis. *Cladistics*, **12**, 221–41.

Nixon, K. C. and Carpenter, J. M. (1996*b*). On consensus, collapsibility, and clade concordance. *Cladistics*, **12**, 305–21.

Nixon, K. C. and Wheeler, Q. D. (1992). Extinction and the origin of species. In *Extinction and phylogeny* (ed. Q. D. Wheeler and M. Novacek), pp. 119–43. Columbia University Press, New York.

Novacek, M. (1992). Fossils as critical data for phylogeny. In *Extinction and phylogeny* (ed. Q. D. Wheeler and M. Novacek), pp. 46–88. Columbia University Press, New York.

Page, R. D. M. (1989). Comments on component-compatibility in historical bio-geography. *Cladistics*, **5**, 167–82.

Page, R. D. M. (1993*a*). On islands of trees and the efficacy of different methods of branch swapping in finding most-parsimonious trees. *Systematic Biology*, **42**, 200–10.

Page, R. D. M. (1993*b*). *COMPONENT version 2.0. Tree comparison software for use with Microsoft Windows. Users Guide*. The Natural History Museum, London.

Patterson, C. (1982). Morphological characters and homology. In *Problems in phyloge-netic reconstruction* (ed. K. A. Joysey and A. E. Friday), pp. 21–74. Academic press, London.

Penny, D. and Hendy, M. D. (1986). Estimating the reliability of evolutionary trees. *Molecular Biology and Evolution*, **3**, 403–17.

Pimentel, R. A. and Riggins, R. (1987). The nature of cladistic data. *Cladistics*, **3**, 275–89.

Pinna, M. C. C. de (1991). Concepts and tests of homology in the cladistic paradigm. *Cladistics*, **7**, 317–38.

Pinna, M. C. C. de (1994). Ontogeny, polarity and rooting. In *Models in phylogeny reconstruction*, Systematics Association Special Volume, No. 52, (ed. R. W. Scotland, D. J. Siebert and D. M. Williams), pp. 157–72. Clarendon Press, Oxford.

Pinna, M. C. C. de (1996). Comparative biology and systematics: some controversies in retrospective. *Journal of Comparative Biology*, **1**, 3–16.

Platnick, N. I. (1979). Philosophy and the transformation of cladistics. *Systematic Zoology*, **28**, 537–46.

Platnick, N. I. (1989). Cladistic and phylogenetic analysis today. In *The hierarchy of life* (ed. B. Fernholm, K. Bremer and H. Jornvall), pp. 17–24. Elsevier Science, Amsterdam.

Platnick, N. I. (1993). Character optimization and weighting: differences between the standard and three-taxon approaches to phylogenetic inference. *Cladistics*, **9**, 267–72.

Platnick, N. I., Coddington, J. A., Forster, R. R. and Griswold, C. E. (1991*a*). Spinneret morphology and the phylogeny of haplogyne spiders (Araneae, Araneo-morphae). *American Museum Novitates*, **3016**, 1–73.

Platnick, N. I., Griswold, C. E. and Coddington, J. A. (1991*b*). On missing entries in cladistic analysis. *Cladistics*, **7**, 337–43.

Platnick, N. I., Humphries, C. J., Nelson, G. J. and Williams, D. M. (1996). Is Farris optimization perfect? Three-taxon statements and multiple branching. *Cladistics*, **12**, 243–52.

Pleijel, F. (1995). On character coding for phylogeny reconstruction. *Cladistics*, **11**, 309–15.

Rieppel, O. (1988). *Fundamentals of comparative biology*. Birkhäuser Verlag, Basel.

Robinson, D. F. and Foulds, L. R. (1981). Comparison of phylogenetic trees. *Mathe-matical Biosciences*, **53**, 131–47.

Rosen, D. E. (1979). Fishes from the uplands and intermontane basins of Guatemala: revisionary studies and comparative geography. *Bulletin of the American Museum of Natural History*, **162**, 276–376.

Sæther, O. A. (1976). Revision of *Hydrobaenus, Trissocladius, Zalutschia, Paratrissocla-dius*, and some related genera (Diptera: Chironomidae). *Bulletin of the Fisheries Research Board of Canada*, **195**, 1–287.

Sankoff, D. and Rousseau, P. (1975). Locating the vertices of a Steiner tree in arbitrary space. *Mathematical Programming*, **9**, 240–6.

Schuh, R. T. and Farris, J. S. (1981). Methods for investigating taxonomic congruence and their application to the Leptopodomorpha. *Systematic Zoology*, **30**, 331–51.

Schuh, R. T. and Polhemus, J. T. (1981). Analysis of taxonomic congruence among morphological, ecological, and biogeographic data sets for the Leptopodomorpha (Hemiptera). *Systematic Zoology*, **29**, 1–26.

Sharkey, M. (1989). A hypothesis-independent method of character weighting for cladistic analysis. *Cladistics*, **5**, 63–86.

Sharkey, M. (1993). Exact indices, criteria to select from minimum length trees. *Cladistics* **9**, 211–22.

Sharkey, M. (1994). Discriminate compatibility measures and the reduction routine. *Systematic Biology*, **43**, 526–42.

Siddall, M. E. (1996). Another monophyly index: revisiting the jackknife. *Cladistics*, **11**, 33–56.

Siebert, D. J. and Williams, D. M. (1997). Book review [Nullius in verba]. *Biological Journal of the Linnean Society*, **60**, 145–6.

Smith, A. B. (1994*a*). *Systematics and the fossil record: documenting evolutionary patterns*. Blackwell Scientific Publications, London.

Smith, A. B. (1994*b*). Rooting molecular trees: problems and strategies. *Biological Journal of the Linnean Society*, **51**, 279–92.

Sokal, R. R. and Rohlf, F. J. (1981). Taxonomic congruence in the Leptopodomorpha re-examined. *Systematic Zoology*, **30**, 309–25.

Stevens, P. F. (1991). Character states, morphological variation, and phylogenetic analysis: a review. *Systematic Botany*, **16**, 553–83.

Stuessy, T. F. (1990). *Plant taxonomy: the systematic evaluation of comparative data*. Columbia University Press, New York.

Sullivan, J. (1996). Combining data with different distributions of among-site variation. *Systematic Biology*, **45**, 375–80.

Suter, S. J. (1994). Cladistic analysis of the living cassiduloids (Echinoidea), and the effects of character ordering and successive approximations weighting. *Zoological Journal of the Linnean Society of London*, **112**, 363–87.

Swofford, D. L. (1991). When are phylogeny estimates from molecular and morphological data incongruent? In *Phylogenetic analysis of DNA sequences* (ed. M. M. Miyamoto and J. Cracraft), pp. 295–333. Oxford University Press, Oxford.

Swofford, D. L. (1993). *PAUP, Phylogenetic analysis using parsimony, version 3.1*. Illinois Natural History Survey, Champaign.

Swofford, D. L. and Begle, D. P. (1993). *User's manual for PAUP, Phylogenetic analysis using parsimony, version 3.1*. Illinois Natural History Survey, Champaign.

Swofford, D. L. and Berlocher, S. H. (1987). Inferring evolutionary trees from gene frequency data under the principle of maximum parsimony. *Systematic Zoology*, **36**, 293–325.

Swofford, D. L. and Olsen, G. J. (1990). Phylogeny reconstruction. In *Molecular systematics* (ed. D. M. Hillis and C. Moritz), pp. 411–501. Sinauer Associates, Sunderland, Massachusetts.

Szumik, C. A. (1996). The higher classification of the order Embioptera: a cladistic analysis. *Cladistics*, **12**, 41–64.

Thiele, K. (1993). The holy grail of the perfect character: the cladistic treatment of morphometric data. *Cladistics*, **9**, 275–304.

Thiele, K. and Ladiges, P.Y. (1988). A cladistic analysis of *Angophora* Cav. (Myrtaceae). *Cladistics*, **4**, 23–42.

Thorpe, R. S. (1984). Coding morphometric characters for constructing distance Wagner networks. *Evolution*, **38**, 244–355.

Turner, H. (1995). Cladistic and biogeographic analyses of *Arytera* Blume and *Mischarytera* gen. nov. (Sapindaceae) with notes on methodology and a full taxonomic revision. *Blumea* (supplement), **9**, 1–230.

Turner, H. and Zandee, R. (1995). The behaviour of Goloboff's tree fitness measure *F*. *Cladistics*, **11**, 57–72.

Vane-Wright, R. I., Schulz, S. and Boppré, M. (1992). The cladistics of *Amauris* butterflies: congruence, consensus and total evidence. *Cladistics*, **8**, 125–38.

Wagner, W. H. (1961). Problems in the classification of ferns. *Recent Advances in Botany*, **1**, 841–4.

Watrous, L. E. and Wheeler, Q. D. (1981). The outgroup comparison method of character analysis. *Systematic Zoology*, **30**, 1–11.

Werdelin, L. (1989). We are not out of the woods yet—a report from a Nobel Symposium. *Cladistics*, **5**, 192–200.

Weston, P. H. (1988). Indirect and direct methods in systematics. In *Ontogeny and systematics* (ed. C. J. Humphries), pp. 25–56. Columbia University Press, New York.

Weston, P. H. (1994). Methods for rooting cladistic trees. In *Models in phylogeny reconstruction*, Systematics Association Special Volume, No. 52, (ed. R. W. Scotland, D. J. Siebert and D. M. Williams), pp. 125–55. Clarendon Press, Oxford.

Wheeler, W. C. and Honeycutt, R. L. (1988). Paired sequence difference in ribosomal RNAs: evolutionary and phylogenetic considerations. *Molecular Biology and Evolution*, **5**, 90–6.

Wiens, J. J. and Reeder, T. W. (1995). Combining data sets with different numbers of taxa for phylogenetic analysis. *Systematic Biology*, **44**, 548–58.

Wiley, E. O. (1981). *Phylogenetics: the theory and practice of phylogenetic systematics*. Wiley Interscience, New York.

Wilkinson, M. (1994). Weights and ranks in numerical phylogenetics. *Cladistics*, **10**, 321–9.

Wilkinson, M. (1995). A comparison of two methods of character construction. *Cladistics*, **11**, 297–308.

Wilkinson, M. and Benton, M. J. (1995). Missing data and rhynchosaur phylogeny. *Historical Biology*, **10**, 137–50.

Williams, D. M. (submitted). An examination of character representation and cladistic analysis: interrelationships of the diatom genus *Fragilarifroma* (Bacillariophyta) and its relatives. *Cladistics*.

Williams, P. L. and Fitch, W. M. (1990). Phylogeny determination using a dynamically weighted parsimony method. In *Molecular evolution: computer analysis of protein and nucleic acid sequences*. Methods in Enzymology, No. 183, (ed. R. F. Doolittle), pp. 615–26. Academic Press, San Diego.

Yeates, D. K. (1992). Why remove autapomorphies? *Cladistics*, **8**, 387–9.

Suggestions for further reading

The number in parentheses after each entry is a cross-reference to the appropriate chapter

Archie, J. W. (1989). A randomization test for phylogenetic information in systematic data. *Systematic Zoology*, **38**, 219–52. (6)

Arnold, E. N. (1981). Estimating phylogenies at low levels. *Zeitschrift für zoologische Systematik und Evolutionsforschung*, **19**, 1–35. (3)

Ax, P. (1987). *The phylogenetic system*. John Wiley, Chichester. (1)

Clark, C. and Curran, D. J. (1986). Outgroup analysis, homoplasy, and global parsimony: a response to Maddison, Donoghue and Maddison. *Systematic Zoology*, **35**, 422–6. (3)

Coddington, J. A. and Scharff, N. (1996). Problems with 'soft' polytomies. *Cladistics*, **12**, 139–45. (4)

Crisci, J. and Stuessy, T. (1980). Determining primitive character states for phylogenetic reconstruction. *Systematic Botany*, **6**, 112–35. (3)

Crowson, R. A. (1970). *Classification and biology*. Atherton, New York. (2)

Davis, J. I., Frohlich, M. W. and Soreng, R. J. (1993). Cladistic characters and cladogram stability. *Systematic Botany*, **18**, 188–96. (6)

De Beer, G. R. (1958). *Embryos and ancestors*, (3rd ed.). Oxford University Press, Oxford. (3)

de Jong, R. (1980). Some tools for evolutionary and phylogenetic studies. *Zeitschrift für zoologische Systematik und Evolutionsforschung*, **18**, 1–23. (3)

Eggleton, P. and Vane-Wright, R. I. (1994). Some principles of phylogenetics and their implications for comparative biology. In *Phylogenetics and ecology* (ed. P. Eggleton and R. I. Vane-Wright), pp. 345–63. Academic Press, London. (1)

Faith, D. P. and Ballard, J. W. O. (1994). Length differences and topology-dependent tests: a response to Källersjö *et al*. *Cladistics*, **10**, 57–64. (6)

Farris, J. S. (1982). Outgroups and parsimony. *Systematic Zoology*, **31**, 328–34. (3)

Farris, J. S. (1986). On the boundaries of phylogenetic systematics. *Cladistics*, **2**, 14–27. (1,3)

Farris, J. S. (1991). Excess homoplasy ratios. *Cladistics*, **7**, 81–91. (6)

Farris, J. S., Källersjö, M., Kluge, A. G. and Bult, C. (1994). Permutations. *Cladistics*, **10**, 65–76. (6)

Felsenstein, J. (1985). Confidence limits on phylogenies: an approach using the bootstrap. *Evolution*, **39**, 783–91. (6)

Felsenstein, J. (1988). Phylogenies from molecular sequences: inference and reliability. *Annual Review of Genetics*, **22**, 521–65. (2,6)

Gift, N. and Stevens, P. F. (1997). Vagaries in the delimitation of character states in quantitative variation—an experimental study. *Systematic Biology*, **46**, 112–25. (2)

Goloboff, P. A. (1991). Random data, homoplasy and information. *Cladistics*, **7**, 395–406. (6)

Gould, S. J. (1977). *Ontogeny and phylogeny*. Belknap Press of Harvard University Press, Cambridge, Massachusetts. (3)

Hauser, D. L. and Presch, W. (1991). The effect of ordered characters on phylogenetic reconstruction. *Cladistics*, **7**, 243–66. (2)

Hillis, D. M. and Bull, J. J. (1993). An empirical test of bootstrapping as a method for assessing confidence in phylogenetic analysis. *Systematic Biology*, **42**, 182–92. (6)

Humphries, C. J. (ed.) (1988). *Ontogeny and Systematics*. Columbia University Press, New York. (3)

Humphries, C. J. and Funk, V. A. (1984). Cladistic methodology. In *Current concepts in plant taxonomy* (ed. V. H. Heywood and D. M. Moore), pp. 323–62. Academic Press, London. (1)

Kitching, I. J. (1992). The determination of character polarity. In *Cladistics: a practical course in systematics*. Systematics Association Publication, No. 10, (ed. P. L. Forey, C. J. Humphries, I. J. Kitching, R. W. Scotland, D. J. Siebert and D. M. Williams), pp. 22–43. Oxford University Press, Oxford. (3)

Le Quesne, W. (1989). Frequency distributions of lengths of possible networks from a data matrix. *Cladistics*, **5**, 395–407. (6)

Lipscomb, D. L. (1992). Parsimony, homology and the analysis of multistate characters. *Cladistics*, **8**, 45–65. (2)

Mabee, P. M. (1993). Phylogenetic interpretation of ontogenetic change: sorting out the actual and artefactual in an empirical case study of centrarchid fishes. *Zoological Journal of the Linnean Society*, **107**, 175–91. (3)

Mabee, P. M. (1996). Reassessing the ontogenetic criterion: a response to Patterson. *Cladistics*, **12**, 169–76. (3)

Maddison, W. P. (1989). Reconstructing character evolution on polytomous cladograms. *Cladistics*, **5**, 365–77. (4)

Maddison, W. P. (1991). Squared-change parsimony reconstructions of ancestral states for continuous-valued characters on a phylogenetic tree. *Systematic Zoology*, **40**, 304–14. (2)

Maddison, W. P. and Maddison, D. R. (1992). *MacClade: analysis of phylogeny and character evolution, version 3.0*. Sinauer Associates, Sunderland, Massachusetts. (2)

Maslin, T. P. (1952). Morphological criteria of phyletic relationships. *Systematic Zoology*, **1**, 49–70. (2)

Nelson, G. J. (1985). Outgroups and ontogeny. *Cladistics*, **1**, 29–45. (3)

Patterson, C. (1980). Cladistics. *Biologist*, **27**, 234–40. (1)

Patterson, C. (ed.) (1988). *Molecules and morphology in evolution: conflict or compromise*. Cambridge University Press, Cambridge. (2)

Patterson, C. (1996). Comments on Mabee's 'Empirical rejection of the ontogenetic polarity criterion'. *Cladistics*, **12**, 147–67. (3)

Penny, D. and Hendy, M. (1985). Testing methods of evolutionary tree construction. *Cladistics*, **1**, 266–72. (6)

Pogue, M. G. and Mickevich. M. F. (1990). Character definitions and character state delineation: the bête noire of phylogenetic inference. *Cladistics*, **6**, 319–61. (2)

Rieppel, O. (1985). Ontogeny and the hierarchy of types. *Cladistics*, **1**, 234–46. (3)

Rogers, J. S. (1984). Deriving phylogenetic trees from allele frequencies. *Systematic Zoology*, **33**, 52–63. (2)

Sæther, O. A. (1986). The myth of objectivity—post-Hennigian deviations. *Cladistics*, **2**, 1–13. (3)

Sanderson, M. J. (1989). Confidence limits on phylogenies: the bootstrap revisited. *Cladistics*, **5**, 113–29. (6)

Sanderson, M. J. (1995). Objections to bootstrapping phylogenies: a critique. *Systematic Biology*, **44**, 299–320. (6)

Sanderson, M. J. and Donoghue, M. J. (1989). Patterns of variation in levels of homoplasy. *Evolution*, **43**, 1781–95. (5)

Sokal, R. R. and Sneath, P. H. A. (1963). *Principles of numerical taxonomy*. W. H. Freeman and Company, San Francisco. (2)

Stevens, P. F. (1980). Evolutionary polarity of character states. *Annual Review of Ecology and Systematics*, **11**, 333–58. (2, 3)

Swofford, D. L. and Maddison, W. P. (1987). Reconstructing ancestral character states using Wagner parsimony. *Mathematical Biosciences*, **87**, 199–229. (2)

Trueman, J. W. H. (1993). Randomization confounded: a response to Carpenter. *Cladistics*, **9**, 101–9. (6)

Trueman, J. W. H. (1996). Permutation tests and outgroups. *Cladistics*, **12**, 253–61. (6)

Wenzel, J. W. (1993). Application of the biogenetic law to behavioral ontogeny: a test using nest architecture in paper wasps. *Journal of Evolutionary Biology*, **6**, 229–47. (3)

Wilkinson, M. (1994). Common cladistic information and its consensus representation: reduced Adams and reduced cladistic consensus trees and profiles. *Systematic Biology*, **43**, 343–68. (5,7)

Wilkinson, M. (1995). Arbitrary resolutions, missing entries and the problem of zero-length branches in parsimony analysis. *Systematic Biology*, **44**, 108–111. (4)

Wilkinson, M. and Benton, M. J. (1996). Sphenodontid phylogeny and the problems of multiple trees. *Philosophical Transactions of the Royal Society of London*, B, **351**, 1–16. (5)

Glossary

Italicized words or expressions in a definition have their own entry in the glossary. 'See also' indicates a cross reference to a related topic, whereas 'Cf.' is a cross reference to an antonym. Synonymous expressions have only a single explanatory entry. 'Also known as' is a cross reference from the main entry to a synonym; 'See' is a cross reference from a synonym to the main entry. Where two or more alternative definitions for a term are provided, then that given first is the accepted usage within the context of this book.

accelerated transformation (ACCTRAN) A procedure for resolving *ambiguous optimization* in which the initial forward character transformations are placed on to the cladogram as close to the *root* as possible. Accelerated transformation accounts for *homoplasy* in terms of reversals to the *plesiomorphic* condition. Also known as *fast transformation*. Cf. *delayed transformation*.

accommodated three-item statement (ATS) A *three-item statement* that fits to a cladogram with a single *step*.

Adams consensus tree A *consensus tree* formed from all of the intersecting sets of internal nodes common to a set of *fundamental cladograms*. Taxa in conflicting positions are relocated to the most inclusive node that they have in common among the fundamental cladograms.

additive coding A method for representing *ordered multistate characters* as a *linked* series of *binary characters*. Cf. *non-additive coding*.

adjacent character Two character states within a *multistate character* are adjacent if they are placed next to each other in a *transformation series*. For example, in the transformation series $0 \leftrightarrow 1 \leftrightarrow 2$, states 0 and 1, and 1 and 2, are adjacent. Cf. *non-adjacent*.

agreement subtree A method of comparing two or more *fundamental cladograms* that shows only the clades and taxa held in common. See also *greatest agreement subtree*.

all-plesiomorphic outgroup An artificial outgroup taxon in which each character is coded with the putative plesiomorphic state as estimated by an *a priori* method of polarization. See also *all-zero outgroup*.

all-zero outgroup An *all-plesiomorphic outgroup* in which all characters are considered to be absent and are coded as zero. See also *all-plesiomorphic outgroup*.

ambiguous optimization The result of *optimization* of a character onto a given cladogram when one sequence of character transformation provides support for a branch, albeit only by *homoplastic* characters or character states, while an alternative sequence produces a *zero-length branch*. The result is an *over-resolved cladogram*.

apomorphy A derived character or character state. Cf. *plesiomorphy*. See also *autapomorphy, homology, synapomorphy*.

autapomorphy A derived character or character state (*apomorphy*) that is restricted to a single *terminal taxon* in a data set. An autapomorphy at a given hierarchical level may be a *synapomorphy* at a less-inclusive level. One form of *uninformative* character.

basal branch See *branch*.

binary character A character that has only two observed states. Binary characters are generally coded 0/1. They can be *directed* or *undirected*, *polarized* or *unpolarized*, but are intrinsically *ordered*. Binary characters cannot be *unordered*. See also *multistate character*.

Biogenetic Law See *recapitulation* (*Haeckelian*).

bootstrap A statistical procedure for achieving a better estimate of the parametric variance of a distribution than the observed sample variance by averaging *pseudoreplicate* variances. The original data set is sampled with replacement to produce a pseudoreplicate of the same dimensions as the original. See also *jackknife*.

branch A line on a cladogram connecting two nodes (*internal branch*), a node and the *root* (*basal branch*) or a node and a *terminal taxon* (*terminal branch*). Also known as an *internode*.

branch length (1) The number of *steps* on a branch. (2) The number of characters that fit to a branch.

branch support See *Bremer support*.

branch-and-bound An *exact* algorithm for cladogram construction. The method begins by constructing a cladogram by means of a *heuristic method*. The length of this cladogram is used as the initial upper bound for an *exhaustive search*. The number of topologies to be examined is then restricted by discarding all partial cladograms whose length exceeds the upper bound. If a complete cladogram is found that is shorter than the upper bound, then the upper bound is reset to this length in order to increase efficiency further.

branch-swapping A procedure for moving clades around a cladogram in an effort to find a more parsimonious topology. See also *nearest-neighbour interchange, subtree pruning and regrafting, tree bisection and reconnection*.

Bremer support The number of extra steps required before a clade is lost from the strict consensus tree of near-minimum length cladograms. Also known as *branch support, clade stability, decay index, length difference*.

Camin–Sokal optimization The *optimization* procedure used for a certain type of *ordered, polarized, directed* character. The costs of the transformations, $0 \rightarrow 1$, $1 \rightarrow 2$, etc., are as in *Wagner optimization*. The costs of all transformations in the opposite direction are treated as infinite, thereby preventing character reversal. Camin–Sokal optimization requires all homoplasy to be accounted for by multiple, parallel transformations.

character (1) A character is an hypothesis of *primary homology* in two or more *terminal taxa* based on original observations of organisms. (2) An observable feature of a organism used to distinguish it from another. See also *character state*.

character analysis A procedure that re-examines the original data in an effort to discover whether any errors in the original coding of characters and scoring of character states have been made, i.e. faulty hypotheses of primary homology or inappropriate character coding.

character congruence See *total evidence*.

character state (1) A scored observation of a feature perceived in the organism(s) chosen to represent a *terminal taxon*. (2) One of two or more alternative manifestations of a *character* (2).

chorological progression, criterion of See *progression rule*.

clade See *monophyletic group*.

clade stability See *Bremer support*.

clade stability index (CSI) The ratio of the minimum number of character deletions from a data matrix required to lose a clade from a cladogram to the total number of informative characters in the data set from which that cladogram was derived.

cladistic consistency The fit of a character to a cladogram in terms of the number of occurrences required to explain the data. Cladistic consistency is usually measured using the *consistency index* or the *rescaled consistency index*. See also *consistent*.

cladistic covariation The degree to which all characters in a data set are explainable by the same cladogram topology.

cladistics A method of classification that groups taxa hierarchically into nested sets and conventionally represents these relationships as a *cladogram*. See also *phylogenetic systematics*.

cladogram A branching diagram specifying hierarchical relationships among taxa based upon *homologies* (*synapomorphies*). A cladogram includes no connotation of ancestry and has no implied time axis. Cf. *phylogenetic tree*.

clique A set of mutually *compatible* characters.

coding The conversion of original observations into a discrete alphanumerical format suitable for cladistic analysis.

combinable components consensus tree A *consensus tree* formed from all the uncontradicted components in a set of *fundamental cladograms*; that is, one that contains all the components found on the respective *strict consensus tree*, plus those components that are uncontradicted by less resolved components within the set of fundamental cladograms. Also known as a *semi-strict consensus tree*.

common pruned tree See *greatest agreement subtree*.

commonality A method of character *polarization* that states that the character state most frequently observed among the *ingroup* taxa is *plesiomorphic*.

compatible character Two characters that do not conflict in the groups that they support are termed compatible. Compatible characters include both those that are *congruent* and those that are *consistent*.

component A group of taxa as determined by the branching pattern of a cladogram. For example, in a group comprising three taxa A, B, and C, where B and C are more closely related to each other than either is to A, there are two components, ABC and BC. ABC is an *uninformative component*, while BC is an *informative component*. Also known as *clade, monophyletic group*.

component information The number of *informative components* in a cladogram. See also *term information*.

compromise tree A consensus tree that includes at least some components that are not present in all of the fundamental cladograms. Cf. *strict consensus tree*. See also *Adams consensus tree, combinable components consensus tree, majority-rule consensus tree, median consensus tree, Nelson consensus tree*.

concordance The degree of agreement between two patterns.

congruence test A test of *secondary homology*. To pass the congruence test, a character must specify the same group on a cladogram as another character. In combination with the *similarity test* and *conjunction test*, the congruence test equates *homology* with *synapomorphy*. See also *conjunction test, similarity test*.

congruent character A character that specifies the same group of taxa as another character. See also *compatible character, consistent character, homologue*.

congruent cladograms A set of cladograms that agree in their topologies.

conjunction test A test of *primary homology*. To pass the conjunction test, two characters hypothesized to be homologous must not coexist in an organism at the same time. See also *congruence test, similarity test*.

consensus method A method for combining the grouping information contained in a set of cladograms for the same taxa into a single topology, the consensus tree. See *Adams consensus tree, combinable components consensus tree, compromise tree, majority-rule consensus tree, median consensus tree, Nelson consensus tree, strict consensus tree*.

consensus tree (1) A branching diagram produced using a *consensus method*. (2) In a restricted sense, a *strict consensus tree*. Cf. *compromise tree*.

consistency index (ci) A measure of the amount of homoplasy in a character relative to a given cladogram. The consistency index is calculated as

Glossary

the ratio of *m*, the minimum number of steps a character can exhibit on any cladogram, to *s*, the minimum number of steps the same character can exhibit on the cladogram in question. See also *ensemble consistency index, retention index.*

consistent character   A character that specifies a subset of the group of taxa specified by another character or a different group of taxa entirely. Also known as *logically consistent*. See also *cladistic consistency, compatible, congruent character.*

constant character   A character for which all taxa in a data set are allocated the same code. One type of *uninformative character.*

constrained, two-step analysis   A method of cladogram construction in which the different states of each character are first organized into *transformation series* and *polarized* using outgroup comparison and known, fixed outgroup relationships. The *synapomorphies* so revealed are used to construct the cladogram. See *Hennigian argumentation; simultaneous, unconstrained analysis.*

continuous character   A character for which potential values are so infinitesimally close that there are potentially no disallowable real numbers, e.g. wing length. See also *discrete.*

convergence   Two characters that pass the *conjunction* test of *homology* but fail both the *similarity* and *congruence* tests are termed convergent. Also known as *homoiologies*. See also *homology, parallelism.*

cost   The number of steps required to account for the transformation of one character state into another on a cladogram.

cost matrix   A square matrix representing the costs of transformation between all states of a character. The values in the upper triangle of the matrix represent the costs in one direction (usually 'forward'), while the values in the lower triangle represent the costs of transformation in the opposite direction.

data decisiveness (DD)   A measure of the degree to which a data matrix is *decisive*. Data decisiveness is calculated as the ratio of $(\bar{S} - S)$ to $(\bar{S} - M)$, where S is the observed length of the most parsimonious cladogram, $\bar{S}$ is the mean length of all possible bifurcating cladograms and M is minimum possible length of a cladogram were there no homoplasy in the data.

decay index   See *Bremer support*.

decisive assignment   The assignment of a unique value to an *internal node* on a cladogram by means of an *optimization* procedure. Cf. *equivocal assignment*.

decisive data   A data set that contains at least one phylogenetically informative character. Decisive data yield cladograms that differ in length among themselves and thus offer reasons for choosing some cladograms in preference to others. Cf. *undecisive data*.

delayed transformation (DELTRAN)   A procedure for resolving *ambiguous optimization* in which the initial forward character transformations are placed on to the cladogram as far from the *root* as possible. Delayed transformation accounts for *homoplasy* in terms of independent gains. Also known as *slow transformation*. Cf. *accelerated transformation*.

dependent character   A character or character state for which the coding depends upon the coding allocated to another character or character state. For example, features of wing venation in insects are dependent upon the presence of a wing.

derived character   See *apomorphy*.

direct method   A method of character *polarization* that can be implemented using only the information available from the taxa under study. Cf. *indirect*.

directed character   A character in which the transformations in one direction cost a different number of steps from the transformations in the opposite direction. For example, in the directed character, $0 \leftrightarrow 1$, the transformation $0 \rightarrow 1$ may cost one step but the reverse transformation, $1 \rightarrow 0$ may cost two steps. *Camin–Sokal* and *Dollo optimizations* both use directed characters. Cf. *undirected*.

direction   The imposition of differential *costs* on the transformation between two character states in one direction relative to the transformation in the opposite direction. See also *order*, *polarity*.

discrete character   A character that can be represented logically by a subset of all possible real numbers, generally only by integers. See also *continuous*.

Dollo optimization   The *optimization* procedure used for a certain type of *ordered*, *directed* character. Characters may be *polarized* or not. The costs of the transformations, $0 \rightarrow 1$, $1 \rightarrow 2$, etc., are as in *Wagner optimization*, but weighted by a high arbitrary value to ensure that each occurs only once on the cladogram.. The costs of all transformations in the opposite direction are as in Wagner optimization. Dollo optimization requires all homoplasy to be accounted for by reversals to more *plesiomorphic conditions*.

doublet Any two consecutive outgroup taxa on a cladogram that share the same state.

ensemble consistency index (CI) A measure of the amount of homoplasy in a data matrix relative to a given cladogram. The ensemble consistency index is calculated as the ratio of M, the minimum number of steps all characters can exhibit on any cladogram, to S, the minimum number of steps they can exhibit on the cladogram in question. See also *consistency index*.

ensemble retention index (RI) A measure of the amount of similarity in a data matrix that can be interpreted as synapomorphy on a given cladogram. The ensemble retention index is calculated is the ratio of $(G - S)$ to $(G - M)$, where G is the greatest number of steps that all characters can exhibit on any cladogram, M is the minimum number of steps all characters can exhibit on any cladogram and S is the minimum number of steps they can exhibit on the cladogram in question. See also *retention index*.

equivocal assignment The assignment of a non-unique value to an *internal node* on a cladogram by means of an *optimization* procedure. Cf. *decisive assignment*.

evolutionary taxonomy A school of systematics that holds the position that overall similarity had to be taken into account and balanced against genealogy in the construction of a classification. Because evolutionary rates among different lineages were known to be highly variable, those taxa that had 'evolved further' (as evidenced by a large number of *autapomorphies*) warranted special recognition. One consequence of this position is an insistence on the retention of *paraphyletic groups* in classifications.

exact algorithm An algorithm for constructing cladograms that is guaranteed to find one or all of the most parsimonious cladograms. Cf. *heuristic algorithm*. See also *branch-and-bound, exhaustive search*.

exhaustive search An *exact* algorithm that examines every possible fully-resolved, unrooted cladogram for the taxa included in that data set in order to find the most parsimonious solution(s).

fast transformation   See *accelerated transformation.*

filter   A procedure applied to characters between the initial discovery phase and the recording of the variation in a data matrix that aims to simplify the *coding* without loss of information.

Fitch optimization   The *optimization* procedure used for *unordered, unpolarized, undirected* characters.

fractional weighting (FW)   The application of differential weights to characters or *three-item statements* in a data set in order to correct for *redundancy.* Most frequently used with reference to *three-item statements analysis.* Cf. *uniform weighting.*

fundamental cladogram   A cladogram produced by direct analysis of data.

gap   A section of a scaled character axis where no observed values occur or where the distance between two consecutive observations exceeds some preconceived value (e.g. one standard deviation about the mean).

gap-coding   A method for recoding *continuous* characters (usually *morphometric* data) as *discrete* characters by the creation or recognition of *gaps.*

gap-weighting   A *gap-coding* method that not only recodes *continuous* characters as *discrete* characters but also maintains the relative sizes of the *gaps* between them by means of *additive* coding.

general character   (1) A character that occurs earlier in an ontogenetic sequence and is considered to specify a more inclusive group. (2) A character that is more frequently observed.

generalized optimization   An optimization procedure in which the costs of transformation are set individually for each character and character state and represented as a *cost matrix.*

geological character precedence, criterion of   See *stratigraphic criterion.*

greatest agreement subtree   The *agreement subtree* found by pruning one or more branches from each of a set of *fundamental cladograms* until a set of identical topologies is obtained. Also known as a *common pruned tree.*

groundplan   A reconstruction of the character states at the *ingroup node.* See also *hypothetical ancestor.*

Hennigian argumentation The first explicit procedure for constructing cladograms, in which the information contained in each character is considered independently. The different states of each character are first organized into *transformation series* and *polarized*, then the *synapomorphies* so revealed are used to construct the cladogram. See also *constrained, two-step analysis*.

heterogeneity Heterogeneity between data sets refers to the statistical difference between the topology and strength of phylogenetic signal contained in two or more data sets coded for the same taxa.

heuristic algorithm An algorithm for constructing cladograms that is not guaranteed to find the most parsimonious solution. Cf. *exact algorithm*. See also *branch-swapping, stepwise addition*.

homoiology See *convergent*.

homologue, homology (1) Two characters that pass the *similarity, conjunction* and *congruence* tests are termed homologous. Also known as *synapomorphy*. (2) Character states that share modifications from another condition, e.g. wings of birds in relation to forelimbs of other tetrapods. See also *convergence, parallelism, primary homology, secondary homology*.

homoplasy (adj. homoplastic) (1) A character that specifies a different and overlapping group of taxa from another character. (2) Any character that is not a *synapomorphy* (*homology*).

hypothesis dependent weighting See *weighting* (*a posteriori*).

hypothesis independent weighting See *weighting* (*a priori*).

hypothetical ancestor A reconstruction of the character states at the *ingroup node* interpreted in terms of a real ancestor of the *ingroup* taxa. See also *groundplan*.

illogical coding See *non-applicable coding*.

implied weighting A procedure for weighting characters according to their fit to a cladogram, assessed as the implied number of extra steps. The fitting function is generally concave, which gives more weight to those characters with least homoplasy. See also *successive* (*approximations character*) *weighting*.

indirect method A method of character *polarization* that requires information from a source external to the taxa under study. Cf. *direct*.

informative component A *component* that includes more than one but less than all of the *terminal taxa* in a data set. Cf. *uninformative component*.

ingroup The group under investigation in a cladistic analysis in order to resolve the relationships of its members. Cf. *outgroup*.

ingroup node The *node* of a cladogram that unites all members of the *ingroup* into a single *clade*. See also *outgroup node*.

interdependency See *linkage* (2).

internal branch See *branch*.

internode See *branch*.

islands of trees Sets of cladograms that can only be reached one from another by *branch-swapping* topologies that are longer than those currently to hand.

jackknife A statistical procedure for achieving a better estimate of the parametric variance of a distribution from small samples than the observed sample variance by averaging *pseudoreplicate* variances. The original data set is sampled without replacement to produce a pseudoreplicate that is smaller than the original. In first-order jackknifing, only a single original observation is excluded; in higher-order jackknifing, more than one original observation is excluded. See also *bootstrap*.

jackknife monophyly index (JMI) The ratio of the sum of $\rho(c_t)$, from $t = 1$ to T, to T, where T is the number of ingroup taxa and $\rho(c_t)$ is the proportion of the most parsimonious cladograms of pseudoreplicate t in which clade c is supported.

jackknife strict consensus tree A consensus tree that includes all components common to or consistent with all members of a set of jackknife pseudoreplicate cladograms.

length (1) The minimum number of character changes (*steps*) required on a cladogram to account for the data. (2) The summed fit of all characters to a cladogram. (3) The total of *accommodated three-item statements* plus twice the number of *non-accommodated three-item statements* on a cladogram.

length difference   See *Bremer support*.

linkage   (1) The condition under which two characters do not represent independent evidence in support of a group. (2) The union of more than two *character states* into a single *multistate* character. Also known as character *interdependency*.

logically consistent   See *consistent*.

Lundberg rooting   A method of rooting that first determines the most parsimonious cladogram for the *ingroup* taxa alone, then, keeping this topology fixed, adds an *all-plesiomorphic outgroup* or *hypothetical ancestor* at the position that gives the least increase in overall length of the cladogram.

majority-rule consensus tree   A *consensus tree* formed from those components that occur in at least 50% of a set of *fundamental cladograms*. The 'majority rule' can also be set to any value greater than 50%.

median consensus tree   A *consensus tree* for which the degree of difference between any pair of *fundamental cladograms* (as measured by a *tree comparison metric*) is smaller than for any other cladogram of the same taxa.

meristic character   A *discrete* character that represents counts of structures expressed as integers, either directly scored into the data matrix or rescaled, e.g. number of digits on the foot of a mammal.

metric tree   A branching diagram on which the length of each *branch* is proportional to the amount of character change that occurs along it. See also *non-metric tree, ultrametric tree*.

midpoint rooting   A method of rooting that places the root at the midpoint of the longest branch or path connecting two taxa.

minimal cladogram   (1) A most parsimonious cladogram. Using the *standard approach*, a minimal cladogram has minimum *length*; using *three-item statements analysis*, a minimal cladogram is one that fits the greatest number of *accommodated three-item statements*. Also known as the *optimal cladogram*. (2) A *strictly supported cladogram*.

monophyly (monophyletic group)   (1) A group is diagnosed as monophyletic by the discovery of *homologies* (*synapomorphies*). Also known as a *clade*. (2) A group that includes a most recent common ancestor plus all and only all of its descendants. See also *component, paraphyly, polyphyly*.

morphometric character A *continuous, quantitative* character derived from measurements of some aspect of the morphology of an organism.

most parsimonious reconstruction See *optimization.*

multistate character A character that has more than two observed states. Multistate characters are generally coded $0/1/2 \ldots n$ and can be *directed* or *undirected, polarized* or *unpolarized,* and *ordered* or *unordered.* See also *binary character.*

nearest-neighbour interchange A method of *branch-swapping* that exchanges a *nearest-neighbour* from one end of an *internal branch* with one from the other end.

nearest neighbours The branches arising from the *nodes* at either end of a particular internal *branch* of a cladogram.

Nelson consensus tree A *consensus tree* formed from the *clique* of mutually *compatible* components that are most replicated in a set of *fundamental cladograms.*

node A point on a cladogram where three or more *branches* meet.

non-accommodated three-item statement (NTS) A *three-item statement* that fits to a cladogram with two *steps.*

non-additive coding A method for representing *unordered multistate* characters as a *linked* series of *binary* characters. Cf. *additive coding.*

non-adjacent character Two character states within a *multistate character* are non-adjacent if there is at least one other character state placed between them in a *transformation series.* For example, in the transformation series $0 \leftrightarrow 1 \leftrightarrow 2$, states 0 and 2 are non-adjacent. Cf. *adjacent.*

non-applicable coding A coding to denote character states that cannot logically be observed in some taxa (e.g. the condition of tooth characters in modern turtles). Computer programs evaluate these as missing data coded as question marks. Also known as *illogical coding.*

non-metric tree A branching diagram on which each of the *branches* is of equal length. See also *metric tree, ultrametric tree.*

ontogenetic criterion A *direct* method of character *polarization* that uses the information derived from *ontogeny* to determine the relative *apomorphy* and *plesiomorphy* of character states found in the *ingroup* taxa.

ontogeny   The pattern of changing features of an organism as development proceeds from zygote to adult.

optimal cladogram   See *minimal cladogram* (1). See also *suboptimal cladogram*.

optimality criterion   The sum total of all constraints to be applied to a character during *optimization* in terms of *order*, *polarity* and *direction*.

optimization   A procedure for reconstructing the most parsimonious sequence of character state change (*most parsimonious reconstruction, MPR*) on a cladogram by minimizing an *optimality criterion*. See also *Camin–Sokal optimization, Dollo optimization, Fitch optimization, generalized optimization, Wagner optimization*.

order   The sequence of character state transformation in a *multistate* character. See also *direction, polarity*.

ordered character   A *multistate* character of which the *order* has been determined. Transformation between any two *adjacent* states costs the same number of steps (usually one, see *direction*), but transformation between two *non-adjacent* states costs the sum of the steps between their implied adjacent states. For example, in the *ordered character*, $0 \leftrightarrow 1 \leftrightarrow 2$, the transformations $0 \leftrightarrow 1$ and $1 \leftrightarrow 2$ each cost the same number of steps but the transformation $0 \leftrightarrow 2$ costs twice as many (i.e. transformation proceeds as if via state 1). *Wagner optimization* uses *ordered characters*. Cf. *unordered*.

outgroup   A taxon used in a cladistic analysis for comparative purposes, usually with respect to character *polarity* determination. Cf. *ingroup*.

outgroup comparison   An *indirect* method of character *polarization* that uses the information on character states in *outgroup* taxa to determine the relative *apomorphy* and *plesiomorphy* of character states found in the *ingroup* taxa.

outgroup node   The *node* on a cladogram that unites the *ingroup* taxa with the first *outgroup* (*sister-group*). See also *ingroup node*.

over-resolved cladogram   A cladogram with *spurious resolution* due to the presence of one or more *zero-length branches*. See also *ambiguous optimization, strictly-supported cladogram*.

parallelism   Two characters that pass both the *similarity test* and *conjunction test* of *homology* but fail the *congruence* test are termed parallelisms. See also *convergence, homology*.

paraphyly (paraphyletic group) (1) A group recognized by *symplesiomorphies*. (2) A group that remains when one or more *components* of a *monophyletic group* are excluded. (3) A group that includes a most recent common ancestor plus only some of its descendants. See also *monophyly*, *polyphyly*.

parsimony The general scientific criterion for choosing among competing hypotheses that states that we should accept the hypothesis that explains the data most simply and efficiently.

partitioned analysis A technique of data analysis in which data from different sources are maintained as distinct data matrices, analysed individually and then the results combined using a *consensus tree* method to extract common groupings. Also known as *taxonomic congruence*.

permutation tail probability (PTP) test A procedure that attempts to assess the degree of *cladistic covariation* in a data set by permuting the states of each character and randomly reassigning them to the *ingroup* taxa in such a way that the proportions of each state are maintained. This process is repeated to produce a large number of such *pseudoreplicate* data sets, which are then analysed cladistically. The PTP is defined as the proportion of all data sets (permuted plus original) that yield cladograms equal to or shorter than those produced from the original data set.

phylogenetic systematics A method of classification that utilizes hypotheses of character transformation to group taxa hierarchically into nested sets and then interprets these relationships as a *phylogenetic tree*. See also *cladistics*.

phylogenetic tree An hypothesis of genealogical relationships among a group of taxa with specific connotations of ancestry and an implied time axis. Cf. *cladogram*.

plesiomorphy (1) An *apomorphy* of a more inclusive hierarchical level than that being considered. (2) An ancestral or primitive character or character state. Cf. also *apomorphy*.

polarity A character or *transformation series* is said to be polarized when the direction of character change or evolution has been specified, thereby determining the relative *plesiomorphy* and *apomorphy* of the characters or character states. See also *direction, order*.

polarization The assignment of *polarity* to a character or *transformation series*.

polymorphic character (1) A character that can show two or more *character states* within the same individual, e.g. alleles. (2) A character that can show two or more states among different individuals of a taxon, e.g. colour forms of some species of butterfly.

polyphyly (polyphyletic group) (1) A group based on *convergent* characters. (2) A group based upon *homoplastic* characters assumed to have been absent in the most recent common ancestor of the group. (3) A group that does not include the most recent common ancestor of all its members. See also *monophyly, paraphyly*.

primary homology An hypothesis of *homology* that has passed the *similarity* and *conjunction* tests, but which has yet to be subjected to the *congruence* test. See also *secondary homology*.

primitive character See *plesiomorphy*.

process partitions A concept developed for *partitioned analysis* in which data are organized into sets satisfying certain restrictive process criteria.

progression rule A method of character *polarization* that states that the character state found in the taxon furthest geographically or ecologically from the ancestral taxon is *apomorphic*. Also known as the criterion of *chorological progression*.

pseudoreplicate An artificial data set produced by permutation of, or re-sampling from, a data set of real observations.

qualitative character A character for which the original observations are in terms of distinct, alternative conditions. Many qualitative characters are actually *filtered* versions of *quantitative* characters.

quantitative character A character for which the original observations are in the form of measurements. See also *meristic character, qualitative character*.

recapitulation (Haeckelian) The view of development that states that ontogeny recapitulates phylogeny and thus the ontogenetic sequence of an organism would be expected to pass through the stages found in the adults of its ancestors. Also known as the *Biogenetic Law*. See also *recapitulation (von Baerian)*.

recapitulation (von Baerian) The view of development that states that two taxa will share the same ontogenetic sequence up to the point that they diverged into separate lineages and thus we would never expect the ontogenetic sequence of an organism to pass through the stages found in the adults of its ancestors. See also *recapitulation* (*Haeckelian*).

redundancy (character) A character that exhibits *linkage* (1) with another character is redundant.

redundancy (three-item statement) A *three-item statement* that is logically implied by the combination of two other three-item statements is redundant. For example, the statement A(BCD) yields the three-item statements: A(BC), A(BD) and A(CD). Combination of any two will recover the original statement, A(BCD), making the third redundant. Redundancy is accounted for by *fractional weighting*.

replicated component A *component* is replicated on two cladograms if it is present in both, irrespective of the internal relationships of its constituent taxa. Thus, if one cladogram includes the clade A(B(CD)) and another cladogram includes the clade A(D(BC)), then the component (BCD) is replicated in both topologies.

rescaled consistency index (rc) The product of the *consistency index* and the *retention index* of a character.

retention index (ri) A measure of the amount of similarity in a character that can be interpreted as synapomorphy on a given cladogram. The retention index is calculated as the ratio of $(g - s)$ to $(g - m)$, where g is the greatest number of steps a character can exhibit on any cladogram, m is the minimum number of steps a character can exhibit on any cladogram and s is the minimum number of steps the same character can exhibit on the cladogram in question. See also *consistency index, ensemble retention index*.

root (1) The basal taxon of a cladogram on which all characters have been *polarized*. (2) The starting point or base of a cladogram.

rooting The process of assigning a *root* to a cladogram.

safe taxonomic reduction A technique used to identify those taxa having a complement of values that will have no influence on the topological relationships in a cladistic analysis.

secondary homology An hypothesis of *homology* that has passed the *similarity, conjunction* and *congruence* tests and is accepted as a *synapomorphy*. See also *primary homology*.

semi-strict consensus tree See *combinable components consensus tree*.

similarity test A test of *primary homology*. To pass the similarity test, two characters must be generally comparable in morphology, anatomy and topographical position. See also *congruence test*, *conjunction test*.

simultaneous analysis See *total evidence*.

simultaneous, unconstrained analysis A method for constructing clado-grams that considers both *outgroup* and *ingroup* taxa together and makes no *a priori* assumptions regarding character *polarity*. See also *constrained, two-step analysis*. Not to be confused with *simultaneous analysis*.

sister-group(s) (1) Two taxa that are more closely related to each other than either is to a third taxon. (2) The taxon that is genealogically most closely related to the *ingroup*.

slow transformation See *delayed transformation*.

spurious resolution Resolution on a cladogram that is not supported unam-biguously by data, but is solely the product of *ambiguous optimization*.

standard approach The method of cladistic analysis that codes the observed features of taxa as *binary* characters and/or *transformation series*, assesses the *optimal cladogram* in terms of *length* and investigates hypotheses of character evolution using *optimization* procedures. Used primarily in contradistinction to *three-item statements analysis*.

step (1) A single gain or loss of a character or a transformation of a *multistate character* on a cladogram. (2) The fit to a node of a clado-gram of an *accommodated three-item statement*.

stepwise addition The sequence by which taxa are added to a developing cladogram during the initial building phase of an analysis.

stratigraphic criterion A method of character *polarization* that states that the character state found in the oldest fossil taxa is *plesiomorphic*. Also known as the criterion of *geological character precedence*.

strict consensus tree A *consensus tree* formed from only those components common to all members of a set of *fundamental cladograms*. In a restricted sense, a strict consensus tree may be considered to be the only *consensus tree* that results from a true consensus. All other consensus methods permit the inclusion of at least some *components* that are not present in all of the fundamental cladograms and thus produce *compromise trees*.

strictly supported cladogram A cladogram from which all *spurious resolution* has been removed and on which all resolved groups are unambiguously supported by data. Also known as a *minimal cladogram* (2). See also *over-resolved cladogram*.

suboptimal cladogram A cladogram that requires more than the minimum number of *steps* to account for the data. Usually interpreted as those cladograms that are one or a few steps longer than the *optimal cladogram*.

subtree pruning and regrafting (SPR) A method of *branch-swapping* that clips off rooted subcladograms from the main cladogram and then re-attaches them in new positions elsewhere on the remnant main cladogram.

successive (approximations character) weighting An iterative procedure for weighting characters *a posteriori* according to their *cladistic consistency*, which is usually measured by the *consistency index* or *rescaled consistency index*. See also *implied weighting*.

symplesiomorphy (1) A *synapomorphy* of a more inclusive hierarchical level than that being considered. (2) The occurrence in two or more taxa of a *monophyletic group* of a *plesiomorphic* character or character state; that is, one that has been inherited from an ancestor more distant than the most-recent common ancestor of the group. *Paraphyletic* groups result from mistaking symplesiomorphies for synapomorphies.

synapomorphy (1) A *secondary homology*. (2) An *apomorphy* that unites two or more taxa into a *monophyletic group*.

taxic approach An approach to cladistic analysis that uses only the distributions of characters among taxa to hypothesize group membership. All other properties of both characters and groups (e.g. *polarity, monophyly, transformation series*) are derived from the resulting cladogram. Cf. *transformational approach*.

taxon A named group of two or more organisms.

taxonomic congruence See *partitioned analysis*.

term See *terminal taxon*.

term information   The term information of a *component* is one less than the number of *terminal taxa* included within that component. The term information of a cladogram is the sum of the term information of all its *informative components*. See also *component information*.

terminal branch   See *branch*.

terminal taxon   A taxon placed at one end of a *terminal branch* on a cladogram. Also known as a *term*.

three-item statement   The concept that two entities (taxa, areas) are more closely related to each other than either is to any other third entity. Also known as a *three-taxon statement*.

three-item statements analysis   A method of cladistic analysis that focuses on the smallest unit of relationship, the *three-item statement*, rather than on characters. The observed features of taxa are coded in terms of the relationships they imply, that is as *three-item statements*, and the *optimal* cladogram is that which maximizes the number of *accommodated three-item statements*.

three-taxon statement   See *three-item statement*.

topology-dependent permutation tail probability (T-PTP) test   A modification of the *permutation tail probability test* that attempts to assess the degree of support for individual clades on a cladogram.

total evidence   A technique of data analysis whereby all data is combined into a single matrix before parsimony analysis to maximize *character congruence* or homology statements. Also known as *simultaneous analysis* or the *character congruence* method.

total support   The sum of the *Bremer support* values of all branches on a cladogram.

total support index   The ratio of *total support* to the length of the most parsimonious cladogram.

transformation series   A series of three or more increasingly apomorphic characters or character states.

transformation series analysis (TSA)   An iterative method of *character analysis* that attempts to bring the *order* of *multistate* characters into conformity with the hierarchy inherent in the rest of the data. TSA begins with the construction of an initial *transformation series* for each multistate character. The data set is then analysed to find the *most parsimonious* cladograms. Any transformation series that conflict with these cladograms are recoded to conform to *adjacent* positions on the cladogram. The data set is then recoded and re-analysed to obtain a new set of most parsimonious cladograms. This process is repeated until both the topologies of the most parsimonious cladograms and the transformation series stabilize.

transformational approach An approach to cladistic analysis that views characters as features of organisms that transform one into another and thus that *polarized transformation series* must be postulated prior to analysis. Cf. *taxic approach*.

tree bisection and reconnection (TBR) A method of *branch-swapping* that clips off subcladograms from the main cladogram and re-roots them before re-attaching them in new positions elsewhere on the remnant main cladogram.

tree comparison metric A measure of the degree of difference between two cladograms. See also *median consensus tree*.

tree dependent weighting See *weighting (a posteriori)*.

tree independent weighting See *weighting (a priori)*.

ultrametric tree A branching diagram on which every *terminal taxon* is the same distance from the *root*. See also *metric tree, non-metric tree*.

undecisive data A data set that includes all possible informative characters in equal numbers so that it is phylogenetically uninformative. Undecisive data yields all possible fully resolved cladograms, which will all be of the same length, and thus offer no reason for choosing some cladograms in preference to others. Cf. *decisive data*.

underlying synapomorphy (1) Close parallelism as a result of common inherited genetic factors causing incomplete synapomorphy. (2) The inherited potential to develop parallel similarities.

undirected character A character in which the transformations in one direction cost the same number of steps as the transformations in the opposite direction. For example, in the undirected character, $0 \leftrightarrow 1$, the transformations $0 \rightarrow 1$ and $1 \rightarrow 0$ each cost the same number of steps. *Wagner optimization* and *Fitch optimization* use undirected characters. Cf. *directed*.

uniform weighting (UW)   The application of equal weights to all characters or *three-item statements* in a data set. Cf. *fractional weighting*.

uninformative character   A character that contains no grouping information relevant to a particular cladistic problem, e.g. *autapomorphies* and *constant characters*.

uninformative component   A *component* that includes either a single *terminal taxon* or all of the taxa in a data set. Cf. *informative component*.

unit discriminate compatibility measure (UDCM)   The complement of the probability of a derived character state being nested with another derived character state, or the probability of a derived character state being exclusive of another derived character state, depending upon the observed pairwise character comparison.

unordered character   A *multistate* character of which the *order* has not been determined. In an unordered character, transformation between any two states, whether *adjacent* or *non-adjacent*, costs the same number of steps (usually one, see *direction*). For example, in the unordered character, $0 \leftrightarrow 1 \leftrightarrow 2$, the transformations $0 \leftrightarrow 1$, $1 \leftrightarrow 2$ and $0 \leftrightarrow 2$ all cost the same number of steps. *Fitch optimization* uses unordered characters. Cf. *ordered*.

unpolarized   A character that has not had its *polarity* determined.

Venn diagram   A graphic representation of a cladogram using internested boxes, circles, ellipses or parentheses to symbolize nodes.

Wagner optimization   The *optimization* procedure used for *ordered, unpolarized, undirected* characters.

weighting (*a posteriori*)   A procedure that applies differential weights to characters following cladogram construction. Also known as *hypothesis dependent* or *tree dependent* weighting.

weighting (*a priori*)   A procedure that applies differential weights to characters prior to cladogram construction. Also known as *hypothesis independent* or *tree independent* weighting.

zero-length branch   A branch on a cladogram that is unsupported by characters.

Appendix: Computer programs

The following is a list of those programs that are mentioned in this book or that implement methods and procedures discussed in the text. More complete listings of phylogenetic computer programs and packages can be found at:
http://evolution.genetics.washington.edu/phylip/software.html
http://www.nhm.ac.uk/hennig/software.html

Farris, J. S. (1988). *Hennig86 version 1.5*. MS-DOS program. Published by the author, Port Jefferson Station, New York.

Goloboff, P. (1996*a*). *PIWE version 2.51*. MS-DOS program. Published by the author, San Miguel de Tucumán, Argentina.

Goloboff, P. (1996*b*). *NONA version 1.50*. MS-DOS program. Published by the author, San Miguel de Tucumán, Argentina.

Maddison, W. P. and Maddison, D. R. (1992). *MacClade 3.01*. Macintosh OS program. Sinauer Associates, Sunderland, Massachusetts.

Nelson, G. J. and Ladiges, P. Y. (1995). *TAX: MSDOS computer programs for systematics*. MS-DOS program. Published by the authors, New York and Melbourne.

Nixon, K. (1992). *CLADOS version 1.2*. MS-DOS program. Cornell University, Ithaca.

Page, R. D. M. (1993). *COMPONENT version 2.0*. MS-DOS program for Windows®. The Natural History Museum, London.

Ramos, T. C. (1996). *Tree Gardener version 1.0*. MS-DOS program. Museu de Zoologia/USP, São Paulo.

Siddall, M. E. (1996). *Random Cladistics, version 4.0.3, Ohio edition*. MS-DOS program. Virginia Institute of Marine Sciences, Gloucester Point.

Swofford, D. L. (1993). *PAUP, Phylogenetic Analysis Using Parsimony, version 3.1*. Macintosh OS program distributed by the Illinois Natural History Survey, Champaign, Illinois.

Index